The Restless Sea

The Restless Sea

EXPLORING THE WORLD BENEATH THE WAVES

ROBERT KUNZIG

W · W · Norton & Company

New York   London

The text of this book is composed in 11/13.5 Fairfield LH Light with the display set in Sanvito Roman and Light
Composition by Binghamton Valley Composition
Manufacturing by The Maple-Vail Book Manufacturing Group
Book design by Margaret M. Wagner
Photograph taken at Big Sur, California, by Bernhard H. Wagner
Cartography by Jacques Chazaud

Library of Congress Cataloging-in-Publication Data
Kunzig, Robert.
The restless sea : exploring the world beneath the waves / Robert Kunzig.
p.   cm.
Includes index.
ISBN 0-393-04562-5
1. Ocean—Popular works.   I. Title.
GC21.K949   1999
551.46—dc21   98-38704
                                                                   CIP

W. W. Norton & Company, Inc., 500 Fifth Avenue, New York, N.Y. 10110
http://www.wwnorton.com

W. W. Norton & Company Ltd., 10 Coptic Street, London WC1A 1PU

1 2 3 4 5 6 7 8 9 0

To my mother and father

Contents

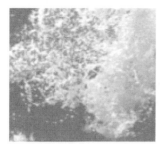

Acknowledgments

I would like to thank:

my friends and colleagues at Discover, especially Marc Zabludoff and Paul Hoffman, for generously encouraging this project with leaves of absence and seaside assignments;

my editor at W. W. Norton, Joe Wisnovsky, for friendship, lunch, and wise guidance, and for a patience that passeth understanding;

my agent, Regula Noetzli, for telling me I could write a book about the ocean, for sticking with me and it, and for agreeing to take full personal responsibility for any errors I may have made;

the many dozens of researchers who gave so generously of their time in interviews, only some of whom are mentioned by name in the text;

the researchers who read portions of the manuscript and helped me make it more accurate, including Colleen Cavanaugh, Sallie Chisholm, Chris Chyba, Fred Grassle, Bill Hamner, Rachel Haymon, Jim Kasting, Ken Macdonald, Larry Madin, Lauren Mullineaux, Phil Richardson, Charles Townes, and George Wetherill;

Sallie Chisholm in particular, for moral support;

Walter Smith, for providing the seafloor map;

Rich Lutz, for taking me to sea on *Atlantis II*, a research vessel;

Tim Ballard, for his friendship and for taking me to sea a long time ago on *Mundeamo*, a 32-foot ketch;

the Marine Biological Laboratory at Woods Hole, not only for repeated use of its wonderful library but also for a generous fellowship at a crucial time;

the Woods Hole Oceanographic Institution, for allowing me on *Atlantis II* and for much other help over the years;

the Scripps Institution of Oceanography;

Tim Appenzeller, Tony Dajer, John Horgan, Jeff Kluger, Anne Lougée, Dan McShea, Sarah McShea, and Marc Zabludoff, for long years of friendship and intellectual exchange;

Dennis Flanagan, friend and mentor;

my parents, sisters, and brothers, for their love, in good times and in tough ones;

my grandmother, for never forgetting to ask whether I had finished Chapter 2;

and Nancy Maynes, my wife, for remaining my wife in spite of this book and for much else too.

New York–Woods Hole–Dijon
1990–1998

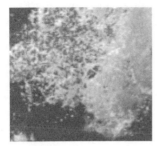

Prologue:

The Sea at Dawn

We put out from Marblehead, my friend, his father, and I, at dawn on a day in early September. The weather seemed promising for a long run down the Maine coast to Portland—cloudless sky and a following breeze. As the morning wore on the breeze freshened, and my friends' 32-foot ketch dashed along with easy grace, drawn by its gaily ballooning spinnaker. We took turns at the helm; we savored the air, the sun-sparkle on the waves, and the feeling of mastery.

There was never a specific moment when the mood changed; it happened imperceptibly. By late afternoon we were surfing down waves that blocked the horizon, at speeds the boat had never reached before. It was exhilarating because it was wild; it was wild because we were overmatched, too much wind driving too much sail. The sky was still clear and the sun was still shining when a strong gust knocked us down, tipping the deck almost to the vertical, dropping the spinnaker in the water and the mast very nearly. As the boat righted itself, we released the spinnaker, belatedly, and it soared off the port side like a kite, with me clenching the line. I could not pull the sail in alone, I realized at that moment; it would pull me over first.

Of course it did not come to that. There were other hairy moments, to be sure, particularly after night fell, but the rest of the story is anticlimax. It is a very ordinary story of the sea. Most of us have similar stories—of rough passages we have endured, of undertows that swept us far from shore, of waves that kicked the breath from us as they flung us on the beach. Most of us have experienced, in some very ordinary way, the danger of the sea, its power to humble us with a casual shrug. And yet we go back to it, again and again. When my friends and I finally dropped anchor in the sheltered waters of Casco Bay, we did not hurry ashore; we made dinner on the boat, and did not want the moment to end.

It is useless to speculate at great length about why the sea has such a hold on us. Its mystery, its seeming infinity must be part of the reason; certainly they help explain why the ocean stirs emotions that can only be described as religious. At some time or another all men and women long to feel small, long for a larger perspective on the trivial noises of day-to-day life. The ocean in its vastness provides that perspective. And along with its vastness comes a visible, enduring wildness. We can raze the Amazonian rain forest if we like; we can settle the Alaskan tundra or the Arabian desert given the right economic incentives. We cannot, to the same extent, tame the ocean, at least not with the tools now available. It retains a unique power to inspire fear, and thus to attract.

This is a special moment in the history of our bond with the sea. Our view of it is very different from what it was just a few decades ago—from what it was in 1951, for example, when Rachel Carson wrote *The Sea Around Us*. Then it seemed ancient and unchanging, tranquil except for its surface layers, a passive receptacle for detritus washed off the continents. Now, thanks to several decades of research, we see the ocean basins as geologically young and constantly in flux, their boundaries continually being rearranged by shifting continents, their depths stirred by giant currents and abyssal storms. The ocean, in short, has become a much more dynamic place; its hallmark is not permanence but ceaseless change.

We ourselves, it is now clear, can change the ocean. Though it is truly untamable it only seems infinite; in fact it is quite small enough for us to muck up. We have always caught as many fish as we could, always treated the ocean as a garbage can, but now there are more than twice as many of us than there were half a century ago, and our technology is more efficient and our garbage more durable. Day after day the newspapers confirm the trend. In the summer of 1997, the "dead zone" around the mouth of the Mississippi River—a fishless region of oxygen-depleted water caused by pollutants washing off the land—was as big as New Jersey. In November 1997, a 70-ton finback whale washed ashore on a beach in northern Spain and died, in spite of earnest rescue attempts. An autopsy revealed that its digestive tract had been clogged by an enormous bolus whose nucleus was 20 kilograms of plastic—shopping bags, trash bags, boat lines, yogurt containers. "The damage is so pervasive," wrote Paul Dayton, a marine ecologist at the Scripps Institution of Oceanography, commenting in early 1998 on the latest overfishing report,

> that it may be impossible ever to know or reconstruct the ecosystem. In fact, each succeeding generation of biologists has markedly different expectations of what is natural, because they study increasingly altered systems that bear less and less resemblance to the former, preexploitation versions. This loss of perspective is accompanied by fewer direct human experiences (or even memories) of once undisturbed systems.

Dayton was writing from direct experience, 25 years of it, of diving in the kelp beds off San Diego. But man's reach extends far beyond coastal waters now. According to one widely credited theory, global warming caused by our burning of fossil fuels could one day realign the great ocean currents themselves—suddenly, and with effects that are not now foreseeable. Not many people are aware that we might have such power. To me it is astonishing.

Our newfound power to damage the ocean lends a special urgency to the work of oceanographers. That work, rather than the environmental damage itself, is what this book is mostly about. I myself am not an oceanographer, or much of a sailor, or an old salt of any kind. I am just a writer who became interested in what science could tell us about the sea, interested enough to spend many evenings, weekends, and vacations looking into the matter. This book, then, is about a long voyage among oceanographers, encountered sometimes on ship or dock, but also in their laboratories and their literature. It is a book of ideas: a portrait of our understanding of the sea, of how we got it and how it is still only inchoate, even as we find ourselves at the dawn of a new era when we are poised to change the sea forever.

There is not much written here about sharks or dolphins or whales or coral reefs; not much either of beaches, waves, or tides. Although all those things are fascinating and important, they are relatively familiar. There is a lot instead about seafloor mountains and stormy deep currents and jellyfish floating in the open sea, far from shore in blue water. I had to limit the subject somehow, and there is so much that is unfamiliar, so much that is unexplored even now.

No president has ever made it a national priority to explore the ocean, the way John Kennedy rallied the nation around a mission to the moon, the way four presidents in a row have committed us to spending tens of billions of dollars on a manned space station of uncertain purpose. It is a paradox I have not quite managed to fathom. Why, when we look for challenges, do we so often tend to look up at the sky rather than at the sea around us? All of us feel the emotional pull of the sea,but few of us feel its intellectual attraction. Many of us still look to the sea for physical adventure, but few of us seek the more serious adventure of discovering something new there. Perhaps the reason is as simple as this: anyone can look up and see the stars from his backyard and dream, but seeing beneath the waves is harder. This book is supposed to give you a glimpse.

That said, to tell the story of the ocean today, one must begin in space, and a long time ago.

The Restless Sea

CHAPTER 1

Beginnings

T hree hundred thousand years after the Big Bang, when the primordial fireball had cooled to a mere 5,000 degrees or so, electrons fell into orbit around protons. The change was sudden, as if a spell had been cast. One instant the electrons were running amok, like children in a schoolyard; the next instant, the universe had expanded and cooled and the negatively charged electrons had slowed just enough to succumb to the positive pull of the protons. The particles lined up two by two. Astronomers speak of this as the moment when the fog cleared, and light could at last course through the universe unimpeded. But to someone interested in Earth's ocean and its origin, the moment has a different significance. A single electron orbiting a single proton is a hydrogen atom; and although individual particles would part and reunite countless times in the eons that followed, it was in that early epoch that the universe acquired its basic supply of hydrogen.

When light stopped being intercepted all the time by matter, matter stopped being pushed around by light. The primeval radiation had been like a wind from all directions, and like the wind it had combed things flat. Freed at last from this smoothing pressure, matter could pursue its own natural tendency to clump. Hydrogen clumped into ever larger clouds, and soon those clouds collapsed, under the force of their own gravitation, to form embryonic galaxies. Within the large clouds, fragments collapsed to form stars. In the cores of stars, history was reversed. Temperatures rose again to what they had been a few minutes after the Big Bang; pressures rose even higher. But now these extraordinary conditions were maintained, not for a transient moment, but for millions of years. There was time now for hydrogen to fuse into two-protoned helium, and for three helium nuclei to fuse into carbon. There was time as well for a carbon nucleus to collect another helium and become oxygen.

As stars evolved, they assembled the elements one after another, on up the periodic table to iron. When a star's core is solid iron, however, fusion stops. Without an internal heat source, the star can no longer fight gravity. It collapses in on itself, compresses itself, rebounds off a 12-mile-wide core of incompressible denseness, and then—all this in milliseconds—blasts itself apart, leaving only that dense cinder behind. Layers of hot star-gas millions of miles thick and laced with millions of years worth of fusion products shoot into space at millions of miles per hour. There they ram into clouds of primeval hydrogen left over from the Big Bang. Thus do interstellar clouds become seeded with heavier elements.

And thus, many billions of years ago, somewhere in our galaxy and probably in a billion other galaxies too, did hydrogen first encounter oxygen.

The recipe for making water was discovered in 1781 by an odd English aristocrat, a man so afraid of women he would fire his own maid if he encountered her on the stairs, so afraid of his peers he might cry out and leave the room if one addressed him, and so afraid of everyone he would take his constitutional at dusk to avoid being seen. According to a contemporary account, Henry Cavendish "was somewhat above the middle size, his body rather thick, and his neck rather short. He stuttered a little. . . ." Another report suggests he smelled bad, but that may be an embellishment. He certainly always wore the same outmoded gray suit. It was men such as Henry Cavendish who established the presumption of wacky eggheadness that scientists to this day must labor under until they prove themselves innocent. Cavendish was the genuine article. He might well have ended up in a public asylum had he not been rich enough to afford his own private one: a large house at Clapham, near London, with a back stair for the maids. There he became instead a remarkable experimental chemist.

By the late eighteenth century, when Cavendish did his great work, three of the four elements into which the ancient Greeks had divided the physical world—earth, air, fire, and water—had lost their elemental status. There were several kinds of earth, several different airs, and fire was not an element but a process. But water was still water, pure and simple. Even Cavendish, who did the experiment that was to topple this intellectual vestige, could never bring himself to see water in any other way.

In 1766, he had isolated what we now know as hydrogen. But Cavendish thought it was phlogiston: a mythical fire essence invented by German chemists, the "oily earth" that was supposedly expelled from a substance when it burned, leaving behind a "stony earth," or ash. Hydrogen burned clean, or so it seemed, and so it seemed a good candidate for pure phlogiston. Eight years later, when Cavendish's compatriot Joseph Priestley isolated a gas that caused a candle to burn furiously and kept a mouse alive for weeks in a bell jar, both he and Cavendish interpreted the discovery in terms of the phlogiston theory. They decided that this exceptionally pure air must be "dephlogisticated air," great for burning things in because it had plenty of room for phlogiston. In fact it was oxygen—great for burning things in because things that burn do not lose phlogiston; they gain oxygen.

Priestley, a carefree empiricist who did experiments almost at random to amuse himself and his friends, soon had the idea of burning Cavendish's phlogiston (hydrogen) in his own dephlogisticated air (oxygen). Holding a lighted candle to the mixture, he succeeded in generating a nice bang. In 1937, the same sort of bang destroyed the *Hindenburg*, the German zeppelin. But Priestley made little of his result. In 1781, Cavendish repeated the experiment more carefully. He noticed that hydrogen did in fact leave a residue when it burned, in the form of a fine mist on the walls of the experimental vessel. He also noticed that the mist was water.

Eventually Cavendish established that if you mixed two parts phlogisticated with one part dephlogisticated air and ignited the lot, then all the air disappeared, and the weight of the water you produced was equal to the original weight of the gases. Cavendish had discovered the right proportions, but he still thought that elemental water was somehow springing whole from each of the two gases.

Two centuries later, when phlogiston and the Greek concept of elements are dead history rather than vital organizing (and eye-blinkering) principles, it is hard to understand his point of view. Antoine-Laurent Lavoisier, cool rationalist of the French enlightenment, was the first to abandon the Greek prejudice. For years he had been quietly undermining the phlogiston theory; for years he had believed that burning is oxidation. When he heard of Cavendish's results, in June 1783, he immediately repeated Cavendish's experiment; tried, in fact, to steal credit for it, for his genius was neither humble nor generous. Yet it was Lavoisier who grasped that the water in Cavendish's vessel had not condensed out of the two gases but had been synthesized from them—from oxygen and the substance that Lavoisier named hydrogen, or "source of water." When Lavoisier triumphantly declared the German phlogiston theory dead, students in Berlin burned him in effigy. Lavoisier responded by staging a ritual oxidation of the book written by Georg Ernst Stahl, chief proponent of phlogiston.

After Lavoisier, water was no longer an element. It was oxidized hydrogen, or burnt hydrogen. Cavendish and Priestley were too old to accept the fact, but the next generation of chemists grew up with it. By the mid–nineteenth century, water was known to be a molecule consisting of two hydrogen atoms and one oxygen: H_2O. With the discovery that water was a compound, not an element, the Greeks had finally been booted out of modern chemistry. Yet a legacy of their thought remained. Although the Greeks did not know the structure of water, they, like the Babylonians and Egyptians before them, appreciated its primordial importance. Homer, in a rare cosmogonical moment in the Iliad, made water the source of deity ("I am off to the ends of the fruitful, teeming earth," Hera says to Aphrodite in Book 14, "to visit Ocean, fountainhead of the gods. . . .") Later, in the sixth century B.C., when Thales of Miletus conceived the first nonreligious cosmogony—the first rational attempt to make sense of the world—he made water its foundation. The world rested on an ocean of water, he said; and all life sprang from water. Thales had the right idea.

As the structure of the atom has been unraveled in the twentieth century, so has the structure of the water molecule. It accounts for water's remarkable properties, which is to say that it accounts—in the sense of being a necessary condition—for the most distinctive properties of our planet, including the fact that it is inhabited.

Water exists because it solves a problem that oxygen has. An oxygen atom has six negatively charged electrons in its outer shell, orbiting the positively charged nucleus, but it has room for and yearns to be filled by two more. Two hydrogen atoms donate the missing electrons and are thereby bound tightly to the oxygen. The eight electrons form four pairs (magnetism keeps each pair together) that separate as far as they can (electrostatic repulsion drives the pairs apart). The result is a four-cornered molecule, a tetrahedron. Projecting from two corners are hydrogen nuclei, that is, individual protons. The other two corners are naked electron pairs. The protons give one side of the water molecule a positive electric charge, while the electrons make the other side negative.

This electric polarization has profound consequences. The naked electron pairs attract their opposite numbers—hydrogen nuclei in other water molecules. The resulting bond is called a hydrogen bond. As long as the forces of chaos—that is, heat—are not extreme, hydrogen bonds link water molecules in a loose association. Each molecule can join hands with a neighbor at the four corners of the pyramid, grabbing a hydrogen at two corners and a pair of oxygen electrons at the other two. The liason is fleeting, though; it is like the passing touch exchanged by dancers in a quadrille, as they constantly change partners. Water molecules change partners billions of times a second.

Their dance, not surprisingly, is not as orderly as a quadrille, and two centuries after Cavendish and Lavoisier, its rules are only beginning to be understood. But its effects are clear enough. It allows water to exist as a liquid on Earth. Without hydrogen bonds, each molecule would go its own way, and water would exist only as a gas. Hydrogen sulfide, the closest molecular analogue to water, does not form hydrogen bonds— sulfur atoms are apparently too big to submit to such constraints—and as a result it does not form a liquid until the temperature drops below −109 degrees Fahrenheit. (That is fortunate, because otherwise our landscape might be dotted with pure pools of rotten-egg smell.) Hydrogen bonds keep water liquid below 212 degrees. They make it possible for Earth to have an ocean.

They also make the ocean Earth's primary distributor of heat. To increase the temperature of a liquid, you have to increase the velocity of its molecules, which in the case of water means you have to overcome the hydrogen-bond restraints. As a result water requires a greater input of heat to raise its temperature by a given amount than does any other liquid. That allows it to store heat and transport it great distances. On Earth, the ocean soaks up heat from the sun in the tropics, where sunlight is plentiful, and transports it to higher latitudes, which can use the warmth. Without the Gulf Stream, for instance, the English would live in a gray and icy land. They have hydrogen bonds to thank (among other things) for their relative prosperity.

Sunlight warms water by causing the molecules to vibrate, straining at their hydrogen bonds, pushing and pulling their neighbors. Water molecules in a liquid can only vibrate in certain modes, however, corresponding to distinct energies, and those energies, in turn, match the energy contained in red light. Thus seawater quickly absorbs red sunlight, whereas it scatters most of the blue. That is why the ocean looks blue to us—when it does not have green or red algae in it, that is, or brown or yellow sediment.

But as sunlight penetrates into the upper layers of the ocean, more and more of all its wavelengths are either absorbed or reflected at each successive layer. By a depth of 500 feet only around 1 percent of it is left. No sunlight at all penetrates below a depth of 3,000 feet or so. Below that depth everything is utterly dark and utterly invisible from the ocean surface.

Because of water, then, we live in a divided world. Thirty percent of it we can see; 70 percent of it we cannot, because it is covered by water, and so we tend to ignore it. It is worth dwelling for a moment on this fact that we take for granted. If we lived on Mars or Venus, our world would contain no insurmountable frontiers. We could get in our sports utility vehicle and drive around the planet on a continuous route, swerving only to avoid mountains and chasms. We could map all parts of it in equal detail—as indeed we have mapped Venus— instead of covering most of it with great swatches of featureless blue. Our world would not end arbitrarily at coastlines. We have the structure of the water molecule to thank, and its propensity to form hydrogen bonds, for our existence on Earth—we could not, in fact, live on Mars or Venus—but also for our monumental ignorance of most of the planet.

The ignorance begins with what sounds like a child's question, but ought to be everyone's first question when standing on the beach: All this water, a billion trillion tons of it, this stuff that makes our planet what it is and makes us who we are—where did it come from? And why did it come here?

A few days before Christmas in 1968, Albert Cheung saw water in space. There was more of it than he or anyone else had ever dreamed he might find. Cheung has since returned to Hong Kong to manage the family business, but at the time he was a graduate student in astronomy at the University of California at Berkeley. His adviser was a physicist named Charles Townes, who had only just switched over to astronomy. Some years before, Townes had conceived the wild notion that molecules, including water, might be plentiful in space.

The idea was wild to most astronomers, anyway. Space is vast, they reasoned, and atoms are small. In a typical cubic foot of our galaxy, there are only a couple of hundred hydrogen atoms, and each atom is just a few billionths of an inch across. Two sparrows released on opposite sides of the Earth and flying randomly have a greater chance of colliding in midair than two atoms have of meeting to form a molecule in most parts of interstellar space. What is more, space is bathed in ultraviolet starlight—the radiation our atmospheric ozone layer still protects us from—which carries enough energy to knock molecules apart. Infrequent births and rapid deaths should result in a relatively small population of molecules between the stars, or so most astronomers figured. "The problem was that astronomers thought they understood interstellar space better than they actually did," Townes recalls, not without satisfaction. "That's why they felt stable molecules couldn't exist and there was no point in looking for them. So they never looked."

Townes, on the other hand, was not overly constrained by knowledge of interstellar space when he first hatched the idea of looking for water and other molecules there. But he had a deep knowledge of how a molecule goes about emitting electromagnetic radiation. As it steps from a higher energy level down to a lower one—each level corresponding to a rotation or a vibration that only that particular molecule is capable of executing—a molecule surrenders energy in the form of radiation. In 1953, Townes had found a way of getting a large population of ammonia molecules to rotate at the same rapid frequency, poised on the brink of a downward energy step. By stimulating them with microwave radiation of precisely that energy, he got them all to take the step at the same time. Thus orchestrated, the molecules delivered themselves of a coherent blast of microwave radiation. Townes had invented the maser—the forerunner and microwave equivalent of the laser. He got the Nobel Prize in 1964.

By then he was bored with masers. The field he had started—the whole idea, he has said, came to him while sitting on a park bench in Washington, D.C., waiting for a restaurant to open for breakfast—had gotten too crowded for him. In 1967, when he moved to Berkeley, he encountered some of the few astronomers who had taken up his earlier suggestion and were looking for molecules in space. Townes decided to get into the act himself, and to start by looking for his old friend, ammonia.

With Cheung and a postdoc named David Rank, Townes built a receiver for the ammonia frequency. An astronomer named Jack Welch helped them install it on the University of California's new radio dish, at Hat Creek in northern California. Pointing the dish toward the center of our galaxy, they pulled in ammonia's 23,870 megahertz signal, the same signal that had coursed through Townes's first maser all those years ago, right away. The ammonia molecules in space were not masing—they were not radiating in lockstep—but they were there. Next Cheung and Townes tuned their receiver to 22,235 megahertz. They pulled in a signal at that frequency, too—the signal that a water molecule makes, floating in space and rotating in its lopsided, rabbit-eared way, with its electrons sliding back and forth like the electrons in a tiny radio antenna.

One evening in December of that same year, 1968, Cheung was back up at Hat Creek. He had the radio dish pointed at the Orion Nebula—the gorgeous glowing cloud, 1,600 light years away, that those of us without telescopes see as the middle "star" in the hunter Orion's sword. Townes was at his home in Berkeley, entertaining the rest of his students and staff at a Christmas party. Sometime during the evening the phone rang: it was Cheung, and he was excited. "It must be raining in Orion," he reported. "There's water everywhere!"

The signal from Orion was far more powerful than what Cheung and Townes had observed coming from the galactic center. The water molecules in Orion were giving off as much energy at a single frequency as the sun emits at all frequencies. It did not take Townes long to figure out what was going on. Fifteen years earlier he had built the first maser in his lab; he and Cheung were using a maser in their receiver to amplify the weak signals they were expecting from space, but in this case the signal coming was amplified already. It was so strong it could have been discovered decades earlier, before Townes's invention, had anyone bothered to point an antenna at Orion and tune it to the right frequency. Cheung and Townes had discovered a water maser in space.

By now it was becoming clear what the astronomers' mistake had been, in assuming all those years that molecules were too unlikely to form in interstellar space and too fragile to withstand the ultraviolet onslaught. Those assumptions were true enough for most parts of our galaxy. But they were not true in scattered dark patches, patches where astronomers had been unable to find any signals at all at the frequencies favored by lone atoms. The dark patches had remained mysterious. There was a patch like that in Orion: It loomed around the visible nebula and behind it, like a larger, hidden truth.

Now it too turned out to be a cloud: a giant cloud of molecular gas, in which matter is packed a million times more densely than it is in ordinary interstellar space. That is still only a hundred-trillionth as dense as the air in our atmosphere. But it is dense enough to allow molecules to form readily; and it is dense enough, given that the cloud is a hundred or more light-years across, to prevent ultraviolet light or any other light from penetrating to its interior. That is why the cloud appears dark. It is also cold, around −400 degrees Fahrenheit. And as astronomers have come to realize in the past three decades, the molecular cloud in Orion and others like it—there are many in our galaxy—are more than just oddities. They are the places where stars are born. Their density and coldness allows gravity to triumph over heat, which is the first requirement if a gas cloud is to collapse in on itself to form a star.

What is more, molecular clouds are lousy with water. In their protected depths, hydrogen and oxygen—hydrogen made in the Big Bang, oxygen made in the interior of stars—are always uniting to form H_2O. The process is not simple. The two flighty hydrogens must first meet on the surface of a frozen dust grain, which holds onto them long enough for them to join hands; and there are several more steps after that. But the process is clearly effective. The microwave signal of rotating water molecules has been detected in just about every molecular cloud. (Unlike ultraviolet or visible light, it is not absorbed by the cloud layers and travels easily all the way to Earth.) The mass of the water in all the clouds in our galaxy is about the mass of a million suns. In most cases, as in Orion, the water is acting as a maser.

It is not clear exactly how interstellar masers work. But it must have something to do with the fact that new stars are forming in the cloud. In the early stages of its life, a star, particularly a massive one, seems to cast off a lot of its mass as a dense wind. Traveling at speeds as high as 250,000 miles per hour, the wind slams into the surrounding gas cloud and compresses it. Water masers are often found at this shock front. The compression of the cloud may bring enough water molecules together in a small enough space to be able to radiate in concert; and radiation from the hot young star may provide the pumping energy needed to excite the molecules and get the maser going.

However they work, water masers are heralds—powerful radio stations, among the most powerful in our galaxy, beaming out the message that a star is born. Four and a half billion years ago, when our own sun emerged from the black, a water maser may have broadcast news of the glad event to the rest of the galaxy. Certainly there would have been no shortage of water in the spinning cloud of gas and dust that gave rise to the sun and its retinue of planets. In such a cloud, water is everywhere. The interesting question is how so much of it ended up in liquid form on just one of the nine planets, the third one from the sun—Earth.

"About ten years ago, while I was on a ride in what used to be called an amusement park," writes George Wetherill, who knows as much as anyone on this Earth about the processes that created it, "a disembodied head appeared above me out of the darkness. Its lips quivered as it intoned: 'So you want to know how the world began? Very well then, I will show you. But remember, you fool, no one asked you to take this journey!'" Wetherill is a modest man. "There really are no bona fide experts on the formation of the Earth," he says. The same is true of the question of how Earth ended up with an ocean.

It was not always thus. Time was when every Great Thinker seemed to be taking a pop at explaining where Earth and the rest of the solar system came from. The first and greatest of these men, at least in the modern scientific era, were Immanuel Kant and Pierre-Simon Marquis de Laplace. (The latter, incidentally, was in Lavoisier's laboratory the day he repeated Cavendish's historic experiment on the composition of water.) Independently of each other, Kant and Laplace conceived essentially the same theory, which came to be called the nebular theory. It attributed the birth of the solar system to a spinning nebula, or cloud, of gas. As the nebula cooled, it shrank, and the sun turned on in its center. As the nebula shrank, it had to spin faster; the rapid spinning produced a centrifugal force that fought gravity and caused the shrinking nebula to leave behind rings of gas, one after another. From those rings the planets condensed.

The nebular theory was neat and pretty. It also explained one of the most conspicuous mysteries of the solar system, which is that all the planets revolve about the sun in the same direction. George W. Wetherill, however, is not overly impressed, at least in this context, by Immanuel Kant and Pierre-Simon Marquis de Laplace. Although at its core the nebular theory expresses a truth that everyone now accepts— that a spinning cloud was the progenitor of the solar system— as far as Wetherill is concerned that is not saying very much. How, exactly, do you get a planet like Earth out of a spinning cloud? "It's amazing how much of this field never got around to making planets," says Wetherill.

There are only two ways of doing it. The first is to have the planet condense from something bigger—a giant gaseous protoplanet. This was an approach many researchers followed until well into the 1960s. But they never spelled out in any detail how the process might have worked, and researchers today generally assume it could not have worked, at least in the case of Earth and the three other rocky planets of the inner solar system. Instead Wetherill and others, following the lead of a Russian planetologist named V. S. Safronov, have adopted the second method of building a planet: assembling it brick by brick, from ever smaller pieces.

The smallest bricks are the tiny dust grains that crystallize out of the gaseous molecules and atoms in all molecular clouds. Just what such grains might look like is an open question, but they include two basic categories of material: rock and ice, that is, solid crystals made of heavy elements and solid crystals made of light ones. In the molecular cloud that hatched our solar system, there must have been rocky grains of pure silicon, grains of iron and other metals, and icy grains of carbon compounds such as methane. Many grains may have had rocky cores with icy coatings. Certainly the cloud contained an abundance of water ice, either as compact crystals or as fluffy snowflakes.

The birth of our solar system began when a dusty cloud, a small fragment of a giant molecular cloud like the one in Orion, somehow became denser than its surroundings and somehow was pushed into collapse, perhaps by the explosion of a nearby star. As the central bulk of the cloud collapsed into a sun, in a hundred thousand years or so, the rest of it began to rotate rapidly, preserving in its motion all the rotational momentum of the original cloud. The centrifugal force of the rotation saved this dust-spiked gas from being swallowed by the newborn star. Instead it drifted down and out, snowing gently into the plane of the sun's equator, where it formed a broad, flat disk.

This disk—a solar nebula, as it is called, with a nod at Kant and Laplace—was denser in the center, near the sun, than it was in its outer reaches. It was also hotter in the center— perhaps as hot as 2,200 degrees Fahrenheit, in the region where Earth now finds itself, whereas out beyond Uranus the temperature would have been more like −400 degrees. And because the central part of the disk was hotter, it was also rockier. Only rocky grains could survive the intense heat; ice, including water ice, would have been vaporized. As a result, the planets that later formed in the inner part of the solar nebula—Mercury, Venus, Earth, and Mars—are mostly rock. The rapid swirling rush of dust in this region may have produced such static electricity that the dust was sometimes rent by thunderclaps and lightning—sound and fury signifying, in this case, the impending birth of a planet.

As the dust grains snowed down into the solar disk, and even more once they were there, they began to collide and stick together. Clumps formed, then bigger clumps, until soon the space that is now occupied by the inner planets was littered with ten billion rocks the size of asteroids, about 5 or 10 miles across. These rocks were large enough to attract one another gravitationally. As they orbited the sun in the same direction, they whipped one another to higher velocities, but whenever they collided they slowed one another down. It was Safronov who first calculated that the net effect of these competing influences would be to keep the relative velocities of the rocks low, and thus the collisions between them gentle—gentle enough that two colliding rocks would often stick together rather than ricochet apart. Within a million years after the sun first coalesced, most of the dust in the inner solar system had settled onto 30 or so planet embryos, the size of Mercury.

Wetherill has calculated what happened next. Now all gentleness faded; now it was survival of the fittest and luckiest embryo. The inner solar system, home today to four planets, was not big enough for 30 Mercuries orbiting on concentric circles. But the embryos did not stay on safe, concentric paths for long. The gravitational attraction among them soon forced the issue. The embryos' orbits became more elliptical, and their paths began to cross—and so they began to collide, not gently, but at speeds of tens of thousands of miles per hour. Only the embryos that started out somewhat larger than the rest could survive such impacts, and they proceeded to grow even larger by absorbing their victims. Within ten to a hundred million years, the planetary barroom brawl had produced two winners, Earth and Venus. Mars and Mercury, which are much smaller, survived only by cowering in corners, well out of the way of the bullies.

———————

This picture of Earth as the survivor of a cosmic free-for-all has, in just the past two or three decades, completely changed our view of the planet's early history, and in particular of how it got an ocean. As long as it was thought that Earth condensed from a giant gaseous protoplanet, or that it grew gently by sweeping up small particles that happened to be in its path, it was also assumed that the primordial Earth was cold. Only gradually did it heat up, in this view, as uranium and other radioactive elements embedded in its rock decayed and released heat; only gradually did the heat drive water vapor, which had attached itself to rocky minerals in the solar nebula, out of Earth's interior. Erupting in great clouds from volcanoes, the water vapor condensed and rained out of the atmosphere. Gradually, over hundreds of millions or even billions of years, the ocean was filled.

The textbooks notwithstanding, it probably did not happen that way. For one thing, the available evidence suggests that over Earth's entire 4.6-billion-year history there has not been enough volcanic activity to cough up an ocean. But in retrospect the whole volcanic "outgassing" scenario rested on a misconception. It is simply not possible for a planet to grow quietly, coolly to maturity. Once the Earth embryo was the size of Mercury, any incoming rock, be it the size of an asteroid or a small planet, would be accelerated to such a terrific velocity by Earth's gravity that on impact it would shatter and melt or even vaporize. Besides releasing a great deal of heat, the explosion would immediately set free any water trapped in the rock. If Earth formed from colliding embryos, it formed not cold but hot, hot as blazes, and, assuming there was water trapped in the rock, it formed wet, right from the beginning.

When one stops to consider just how much heat and how much water might have been released by the torrent of planetoids that pelted Earth during its frenetic adolescence, one arrives at an astonishing possibility, first pointed out in 1986 by Takafumi Matsui and Yutaka Abe of the University of Tokyo. The cloud of water vapor would have lain like a thick blanket over the young planet, and like a blanket it would have trapped heat—so much heat that the water would have become steam. And as the steam got thicker and thicker with each fresh impact, almost none of the heat released by the exploding projectiles would have been able to escape into space. The temperature of the planet surface would have risen inexorably, until finally something gave: until Earth's entire surface melted. Our planet's first ocean, in this scenario, was an ocean of liquid rock.

The magma ocean would have prevented the steam from thickening and the temperature from rising indefinitely. As incoming rocks released steam, some of it dissolved in the magma. At a temperature of 2,300 degrees Fahrenheit, according to Matsui and Abe, the two processes would have been in balance: the same amount of steam was being released into the atmosphere as was being drawn out by the magma. At that point the atmospheric pressure would have been 100 times what it is today. The weight of the steam atmosphere would have been around a billion trillion tons.

What would this primordial Hades have been like? Perhaps like a pair of Rothko's color fields, painted large: A bright-red ocean, the color of the lava that flows out of Kilauea, on Hawaii, under a perpetual gray fog. A boundless ocean of churning, boiling rock, stirred from time to time by the splashdown of a minor planet, which would radiate tsunamis of magma in all directions; a thick, viscous ocean with no land, no landmark at all to catch the eye, even if the eye had been able to penetrate more than a few feet through the dense, crushing fog. A boundless shimmering bright-red ocean of rock below, and above, like some giant inflatable dome, propped up precariously by the heat of sporadically exploding projectiles, a whole scalding ocean of water. The situation could not last. Eventually the rain of projectiles had to taper off, and the surface of Earth had to cool. Eventually the rain of water had to begin.

It would have been a cooling rain by the standards of the day, but its temperature would have been 600 degrees, because that (and not 212 degrees) is the temperature at which water starts to condense to a liquid under 100 atmospheres of pressure. The rain, once it began, would have been implacable. It would not have been a rain as we know rain, an event that leaves no permanent mark on the atmosphere; it would have been the collapse of the atmosphere. It would have lasted much longer than 40 days and nights—millennia, perhaps, during which time virtually all the water in the atmosphere fell to Earth. The amount of water a steam atmosphere can support, Matsui and Abe calculate, is very close to the amount of water in the ocean today—1.4 billion trillion tons—an encouraging coincidence for their scenario. That first ocean of water would have been boiling hot.

It may or may not be the same ocean we have today. One factor not taken into account in the Matsui and Abe scenario is the dual nature of the enormous projectiles striking Earth: they gave our planet water, but the largest ones also took water away, blasting it into space. In particular, it is widely believed today that Earth was struck, very early in its history but at a time when it was almost full-grown, by a planet embryo the size of Mars, or maybe even twice the size. That impact melted our planet right through and perhaps vaporized part of it. The incoming embryo too was vaporized, and most of it, all but its heavy core of iron (which sank to the center of Earth and merged with Earth's own core), rebounded back into space. Earth's gravity was so strong by then, however, that the debris

of the embryo could not escape completely. Instead the fragments quickly reaccreted, in as little as a year, to form a solid
body in orbit around our planet: the body we know as the
moon.

Such an impact would certainly have stripped the planet of
its steam atmosphere, if that was still around at the time. If
the atmosphere had already collapsed into an ocean, that
ocean would have been boiled away. Thus the ocean we have
today must, if this scenario is true, have been born after the
moon. The same processes that created the first steam atmosphere and the first ocean, though, could easily have endowed
the planet with a second and more lasting one—assuming, that
is, that Earth was still accreting enough rock from space, and
that the rock contained enough water.

That assumption may or may not be correct. It is not clear
that the rocks that made the solid planet were also wet enough
to give it an ocean. Many researchers think not. They think
Earth must have gotten a lot of its water from somewhere else.

Most human beings who look at the ocean are impressed by
the amount of water they see. When they look at pictures of
Earth from space, they are impressed again: it is the Blue
Planet, the Water Planet. James Pollack looked at Earth and
the amount of water on it, and he said: "It's just an incredible
depletion. The Earth has nothing close to what it could have
ended up with." And he looked at the solar system as a whole
and saw a strange paradox: "On the one hand, the terrestrial
planets and particularly the Earth are located just in the right
temperature zone, at the right distance from the sun, so you
could have oceans of water—water in the liquid phase rather
than in the solid phase. But on the other hand, the region that
is really just enormously rich in all volatiles, including water,
is the outer part of the solar system, beginning somewhere in
the outer asteroid belt and continuing all the way out."

Pollack, who died in 1994, was a planetary scientist at NASA's Ames Research Center in California. He had started his career in the early 1960s in the vicinity of Venus, and then moved steadily outward, following the trail of NASA probes—the Pioneers and Voyagers—in their grand missions of discovery. Those probes revealed what the giant outer planets (Jupiter, Saturn, Uranus, and Neptune; Pluto has not been visited yet) are made of, along with the composition of their many moons. A typical moon, it seems, is half rock and half ice, mostly water ice. The giant planets themselves consist of cores of rock and ice enveloped in thick, dense atmospheres of hydrogen and helium gas. All this makes sense: in the outer part of the solar nebula it would have been cold enough not only for rock to condense, as in the inner nebula, but also for water and methane to freeze into solid ice; but it was not cold enough to condense hydrogen and helium, nature's flightiest atoms.

If one considers, though, as Pollack did at length, the question of how exactly the giant planets might have formed, one arrives at a problem. The cores of rock and ice could have emerged from the same brick-by-brick process of accretion that Wetherill has outlined for the inner planets. The nascent giants could then, by dint of their powerful gravity, have cloaked themselves in hydrogen and helium drawn from the solar nebula. But the problem is this: they had to be quick about it, because the boisterous young sun would have cleared most of the gas out of the nebula within the first ten million years. The rock-ice cores had to reach full strength before then. Yet if one takes their present mass and spreads it around the enormous length of their far-flung orbits, it becomes hard to see how the bricks could have come together in time.

Pollack came up with a solution: more bricks. If planetesimals had been packed much more tightly in the outer solar nebula, perhaps ten times as many of them as ended up in the cores of the giant planets, the planets could have grown large enough fast enough to trap enough gas. But that raises a new problem: how to get rid of the other 90 percent of the plane-tesimals, the wasted bricks that were not incorporated in the finished buildings. That is where Earth's ocean comes in, and Halley's comet.

In 1986, when a flotilla of European, Russian, and Japanese spacecraft flew by Halley's comet, they confirmed what had long been suspected about its composition. It is primarily a mixture of rock and water ice in roughly equal proportions—a dirty snowball or snowy dirtball, depending on your point of view. In other words, it is about the same composition as an outer-planet moon. Comets today are very different from outer-planet moons. They course through the solar system on crazy orbits, streaking toward Earth from all directions. The long-period ones like Halley apparently come from a spherical, shell-like cloud that surrounds the entire solar system at a distance well beyond Pluto. The comets we see are ones that have been knocked out of this Oort Cloud (it is named after the Dutch astronomer, Jan Oort, who first imagined it) by a passing star. And yet the similarity in composition between Halley and the outer-planet moons cannot be a coincidence: Halley and most of the other long-period comets are now thought to have coalesced in the region around Uranus and Neptune.

As Pollack saw it, comets are failed planetesimals, discarded into the solar system's outer precincts after they had outlived their usefulness in helping to build the giant planets. Once Uranus, say, had reached its present mass, nearly 15 times that of Earth, its relation to a planetesimal would have been that of Hercules to a discus. A planetesimal passing near Ura-nus, but not near enough to fall in, would be gripped by the planet's powerful gravity and acccelerated until its velocity was high enough to escape. At that point it would be flung off on a different orbit. It might streak off toward the outer fringes of the solar system and take up residence in the Oort Cloud.

Or it might be hurled in the other direction, toward the inner solar system. There it would have crashed in on the brawl of rock that was just then being decided in favor of Earth and its three sister planets.

Jupiter and Saturn, more Herculean even than Uranus, would have been ejecting their own leftover snowballs at the same time, scattering them all over the field of the young solar system—and in the case of Jupiter, even right out of the stadium. The tumult that accompanied the birth of the giant planets thus provides the link that resolves Pollack's paradox: the link between the outer solar system, where water is plentiful, and the inner solar system, where it can exist as a liquid. Halley's comet may contain hundreds of billions of tons of water. A typical outer-planet moon contains thousands of times that amount. If just a small percentage of the planetesimals that originally condensed in the outer solar system had struck Earth, they could have endowed it with an ocean of water.

If most of them struck while the rocky part of Earth was still coalescing, the result would have been much as if the rocks themselves had spit out the water, as Matsui and Abe envisioned. The same massive steam atmosphere would have formed, and the same magma ocean, until the steam collapsed into an ocean of liquid water. It is also possible, though, that the rain of icy planetesimals was drawn out over hundreds of millions of years. There is no doubt that Earth was being peppered by projectiles of some kind for the first 800 million years or so of its existence; the ages of craters on the moon, as inferred from rocks retrieved by the Apollo astronauts, indicate that the moon was being heavily bombarded by the flying debris of planet formation until 3.8 billion years ago. Earth, being so nearby, must have experienced the same prolonged bombardment. Many of those projectiles may have been icy planetesimals, arriving from the outer solar system by circuitous routes, perhaps after brief careers as comets. Instead of a single massive steam atmosphere, Earth may have had a series of thinner and more transient ones, as each exploding comet vaporized part of the ocean and added its own mushroom cloud of water to the expanding pool.

Is our ocean pure comet water then? One recent observation suggests not. In addition to the Halley flyby in 1986, Earth itself had close encounters in 1996 and 1997 with two other comets, Hyakutake and Hale-Bopp. In all three comets, astronomers succeeded in measuring the ratio of deuterium, hydrogen's heavy isotope, to ordinary hydrogen. Seawater has a higher ratio than the solar system average; it is richer in deuterium. This is one reason Pollack and others have argued the ocean could not have been squeezed out of the ordinary rocks that made the solid Earth. But the three comets all contain twice as much deuterium even than seawater. Thus our ocean could not have been made purely from objects like them. Three comets are a small sample of the trillion or so in the Oort Cloud, but that is more data than students of Earth's origins usually get. The prudent conclusion at the moment is that comets provided no more than half the water in our ocean, and perhaps much less. The comet water had to have been cut with something else.

That something else might still have come from the outer solar system though, and in the way that Pollack envisioned. The meteorites called carbonaceous chondrites, which are believed to have fallen to Earth from the outer half of the asteroid belt between Mars and Jupiter, are extremely wet as rocks go. Perhaps early in the planet's history Jupiter was firing volleys of those wet rocks at us along with the dirty snowballs.

No one knows. It is not possible now to say definitively where the ocean comes from—from the rocks we now walk on, from an early rain of icy planetesimals that leavened the accreting rocks, or from a late rain of comets that added a veneer of water to an already assembled planet. The truth probably contains elements of all three scenarios. But the common thread, and the great change from the view that prevailed just a few decades ago, is that Earth received its water from large objects slamming into it at high speed—from the same violent process that gave rise to the planet itself, and indeed to all planets.

If not all those large objects were comets, then certainly at least some of them were. And that means at least some of the water molecules in our ocean are truly original: they were made before Earth itself, even before the sun, in interstellar space; they were frozen into solid grains of ice in the cold outer reaches of the solar nebula; and finally they were brought to Earth by the strange solar system travelers that much later, in the strange human mind, would become symbols of such portent. Nor is water the only relic of the mother cloud that comets brought us. The most exciting news from Halley, Hyakutake, and Hale-Bopp was not that comets contain a lot of water, which no one doubted. It was that comets also contain a great deal of organic matter, that is, compounds of carbon and hydrogen, sometimes including nitrogen or oxygen as well. About a quarter of Halley's mass was made up of organic compounds. Notable among them were formaldehyde (H_2CO) and hydrogen cyanide (HCN). Both of these molecules have been seen in interstellar clouds. They also happen to be two of the most important chemical precursors of the more complex molecules—proteins, nucleic acids such as DNA, and lipids—that may properly be called the molecules of life.

At the southern tip of Shark Bay, on the hot and arid west coast of Australia, is a shallow, salty lagoon called Hamelin Pool. When the tide is out, the bottom of the lagoon is revealed. It is an eery sight. The sand is studded with large greenish-brown lumps, shaped a bit like cauliflowers, a couple of feet tall and wide. These lumps are alive. The top layer of each one is a living mat of cyanobacteria, woven together in a web of filamentous colonies. Like algae and other plants, cyanobacteria—formerly called blue-green algae—have the ability to capture the energy of sunlight through photosynthesis. As the tide washes over the lumps in Hamelin Pool, the bacteria extract carbon dioxide from the water and, using solar energy, construct more complex organic compounds from it. In the process they precipitate calcium carbonate, the stuff of limestone. Every now and then a surge of storm-washed sediment becomes trapped in the web of bacterial filaments. The bacteria grow upward, through the mud and toward the light, leaving behind a hard reef made of thin, alternating pancakes of white limestone and dark sediment.

The lumpish reefs of Shark Bay are living fossils—the closest thing we know to the oldest known organisms. From the time life began until between two billion and one and a half billion years ago, Earth was in the exclusive custody of bacteria: single cells that lack the nucleus found in all higher organisms from amoebas to *Homo sapiens*. Even after nucleated cells emerged, mat-forming, reef-building cyanobacteria continued to dominate the planet. They lived everywhere, in the open ocean and on desert continents, which were still barren of all the more obvious forms of life; they lived from the high latitudes to the tropics and from icy freshwater lakes to hot hypersaline lagoons like Shark Bay. The reign of the cyanobacteria ended only around a billion years ago, when Earth was already three and a half billion years old. That is when multicellular animals emerged and began to ride a novel life strategy, founded on eating cyanobacteria, to success. Today, although cyanobacteria remain common in the ocean, the reef-building ones survive primarily in places like Shark Bay, where conditions are hostile enough (the water there is nearly twice as salty as the open ocean, and the bacterial mats are often left out to dry in the blazing sun) to discourage potential grazers.

When did the reign of the bacteria begin? That is, when did life begin? Five hundred miles northeast of Shark Bay, in the desert of northwestern Australia, there is a rock formation called the Warrawoona Group that contains the oldest fossil reefs on the planet. The remains of life have long since decayed or been squeezed out of these stromatolites, as they are called; but their distinctively layered sediments and their bulbous form are a clear indication of their origin. And in a stratum of the Warrawoona Group called the Apex chert, J. William Schopf of the University of California at Los Angeles has found the reef builders themselves: fossilized filaments of cells, a dozen or more in a chain, each no more than a few micrometers wide. These cells are 3.5 billion years old. They are the oldest fossils that have yet been found.

But they are certainly not fossils of the first organisms. Schopf has found at least 11 different types of cells, some resembling ordinary bacteria and others cyanobacteria, in the Apex chert. Such diversity indicates that evolution was already well underway 3.5 billion years ago and that the origin of life lies farther back in time. A 3.8-billion-year-old rock formation at Isua, on the west coast of Greenland, offers the next milepost: it includes no distinct fossils but does contain a kind of chemical footprint (in the form of an unusual ratio of carbon isotopes) that may indicate living cells were there, then. All in all, it seems likely that life began at least 3.8 billion years ago. It could not have gained a permanent foothold much before then, however, assuming the first organisms required sunlight and thus had to live near the surface. Until 3.8 billion years ago, when the heavy bombardment of planetary debris ceased, the surface of the ocean was probably being boiled and sterilized again and again as each new planetesimal plunged into it and kicked up a blanket of steam.

"A little more than 3.8 billion years ago," then, is as good an answer as any to the question of when life began on Earth. The answer to the question of how remains speculative. There is no standard theory; there is instead a confusion of conflicting or incommensurable theories that attack the problem from different angles. This is a change from 40 years ago. Then, Stanley Miller and Harold Urey had just completed their famous laboratory simulation of "primordial soup" at the University of Chicago. When Miller and Urey let electric sparks course like lightning through an "atmosphere" of methane, ammonia, and hydrogen, which circulated above an "ocean" of boiling water, they found that a reddish substance, rich in amino acids, accumulated in their glass apparatus. Amino acids, when strung together in long folded chains, form proteins, and proteins are the workhorses of the living cell. From the spontaneous synthesis of amino acids to the spontaneous origin of life on the primitive Earth did not seem such a long way to go.

That early optimism has proved profoundly mistaken, for at least two reasons. The first is simply that it is, in fact, a long way from amino acids to life. The hardest part about creating life is not making the amino acids that go into proteins; or the sugars, phosphates, and bases that go into DNA, which carries the cell's genetic blueprint; or the lipids that form its protective membrane. No, the hardest part about creating life, like creating a planet, is not making the bricks: it is assembling them into a finished structure. That is what all the theories that have emerged since the Miller-Urey experiment are primarily about; and the conflict among them shows no signs of being resolved soon.

Furthermore, in recent years even the fundamental premise of that landmark experiment has been called into question. Today most researchers who study early Earth do not believe that its atmosphere was primarily methane and ammonia— that it was a strongly reducing atmosphere, where "reducing" means hydrogen-rich. Methane (CH_4) and ammonia (NH_3) are both comparatively fragile molecules that might easily have been broken apart by the ultraviolet sunlight that bathed the young Earth, which had not yet evolved an ozone shield. More important, the whole idea that Earth was hot to begin with, as a result of its violent birth, implies that its early atmosphere was rich in carbon dioxide rather than methane. That is the form in which carbon would be released by exploding planetesimals.

The bottom line is that the early atmosphere is not likely to have been a giant Miller-Urey experiment; it would have been mostly nitrogen and carbon dioxide. In such an atmosphere it is indeed hard to make the molecular bricks of life, let alone a living organism. It is hard even to make the chemical precursors of the bricks. The most important precursors are formaldehyde and hydrogen cyanide, both of which, curiously enough, are poisonous to living cells today. But brought together in the presence of water, they react to produce amino acids. Hydrogen cyanide also transforms into the chemical bases that are the steps in DNA's spiral staircase and that carry its genetic information, whereas formaldehyde reacts to form the sugars that are DNA's backbone. Formaldehyde and hydrogen cyanide, then, seem to be essential stages on the chemical road to life—and hydrogen cyanide especially cannot be made in great quantities in a carbon-dioxide atmosphere. Both compounds, however, are abundant in Halley, Hyakutake, and Hale-Bopp. Presumably they are in other comets as well.

Here then is an elegant solution to the dilemma. The dilemma is that the old view of how life began conflicts with the new view of how Earth began and how it acquired an ocean. The solution, perhaps, is to deliver the organic precursors of life with the same vehicles that almost certainly helped create the ocean: icy comets. Christopher Chyba, now at the University of Arizona, has calculated that over the course of Earth history, comets have delivered an amount of organic matter to the planet that is nearly a million times its present biomass—the total mass of all living things. Most of the organic matter would have arrived during the heavy bombardment that ended 3.8 billion years ago. And most of the organic compounds would have been destroyed, torn to atoms, by the tremendous heat released by each impact. But some of them could have survived. Medium-size comets, larger than a couple of hundred meters across, tend to disintegrate in the atmosphere; instead of reaching the surface as a cannonball, they might arrive as a slower rain of shrapnel, with some of their organics unharmed.

More important, comets as large as Halley scatter organic matter on us all the time. Their long tails consist in part of microscopic dust particles that are rich in organics. As Earth passes through one of those dust trails, some of the particles burn up in the atmosphere and are seen as meteor showers, but others float gently to the surface. By one recent estimate these particles are still supplying Earth with 300 tons of organic carbon a year. During the heavy bombardment, says Chyba, when comets were rife, they may have delivered anywhere from 100 thousand to 10 million tons a year of organic matter in this way—streaking past the barren planet like crop-dusting aircraft, but scattering seed rather than poison.

It is an appealing image. But as with most images of our beginnings, there can be no certainty about it. Only one thing seems reasonably sure: the origin of life on Earth required water. Living organisms today are mostly water. By virtue of its electrically polarized design—positively charged hydrogen and negatively charged oxygen at opposite sides of the molecule, both ready and able to grab onto other things—water is the universal solvent, the stuff that brings atoms and molecules together and also tears them apart. Water carries nutrients into the living cell, and it removes wastes. All biochemical reactions take place in water. The chemical evolution that culminated in the first living, self-replicating organism had to take place in water as well.

In our solar system Earth is the only planet with liquid water on its surface (at present, anyway) and, as far as we can tell, the only planet with life. With rock and ice flying every which way in the solar nebula, Venus, Earth, and Mars would have had a common endowment of water. But only Earth managed to hold onto it in liquid form. Why? In the case of Venus the answer is fairly simple: it is too close to the sun. It may never have cooled enough to allow the water in its atmosphere to condense into an ocean; if it did have an ocean early in its history, it never cooled enough to hold onto that ocean permanently. Like the young Earth, the young Venus may have had a thick atmosphere of carbon dioxide and water vapor, both of them effective at trapping the heat of the sun. But the Venusian greenhouse, being less than three-quarters as far from the sun as Earth, received nearly twice as much sunlight. So it stayed hotter. The heat allowed water vapor to rise into the upper atmosphere, where it was torn apart by solar radiation. The hydrogen then escaped into space, taking with it, forever, the planet's potential for water. Today the Venusian atmosphere is 96 percent carbon dioxide, and it is 90 times as thick as Earth's. The temperature on its surface averages around 850 degrees Fahrenheit.

Mars is an altogether different case. It is half again as far from the sun as Earth is, but that is still close enough for liquid water to exist. The network of river valleys that cut through its ancient southern highlands prove that Mars did have liquid water early in its history. It may even have had primitive life: in 1996 NASA scientists claimed to have found microscopic fossils in a meteorite that fell to Earth from Mars. That claim is questionable and controversial, but everyone agrees that now, anyway, Mars is bone dry and dead. The reason is not that Mars is too far from the sun, Pollack and his colleagues, James Kasting of Pennsylvania State University and Owen Toon of NASA Ames, concluded some years ago. It is that Mars is too small.

The rains that filled the early Martian ocean, they propose, swept carbon dioxide out of the atmosphere as well, locking it up in rock as calcium carbonate. For hundreds of millions or even billions of years, the interior of Mars remained hot, and erupting volcanos returned carbon dioxide to the atmosphere. The carbon dioxide—some of it perhaps in the form of clouds, and assisted perhaps by a trace of methane—kept the surface warm enough for liquid water. But eventually the volcanos were stilled. Having a diameter only half that of Earth, Mars absorbed much less heat from the collisions that formed it. Having a volume less than one-sixth that of Earth, it generated proportionately less heat from the decay of radioactive elements embedded in the rock. As a result, Mars simply ran out of steam. The last carbon dioxide taken from its atmosphere never got returned. The Martian atmosphere today is less than a hundredth as thick as Earth's, and the planet is frozen solid. Whatever water it has retained is locked up in permafrost and in polar ice caps.

Earth has an ocean today because its interior is still hot. The heat drives volcanoes that return carbon dioxide to the atmosphere and thus keep our greenhouse at a friendly temperature. All around the rim of the Pacific, volcanoes bubble with activity; all down the center of the world ocean, hidden from view by miles of water, a continuous range of volcanoes erupts continually but sporadically, in a syncopated firing line. The mechanism that orchestrates all this activity was scarcely dreamed of 50 years ago. Then people regarded the ocean basins as ancient features and as unchanging ones—tedious deserts, uncharted but hardly worth charting, vacant spaces disturbed only by endless rains of sediment. If the ocean basins really were like that, it is now clear, we would not have oceans at all. There are oceans today on Earth, and there is life today on Earth, because the ocean floor itself is young and moving and constantly changing and, in that sense, alive.

CHAPTER 2

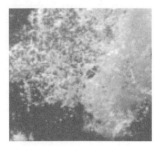

The Seafloor

Moves

If only there were no water! If only the ocean basins were drained, just for a month or a week, or even a day. Then we could stand like Balboa atop a Sierran peak, or better yet at the edge of the continental shelf, where the land really stops, and see a truly New World where the Spaniard saw only blank Pacific. We could fly low over the route Columbus took, marveling at all he passed over in ignorance. Retracing his route on foot, we could hike across some of Earth's flattest and most expansive plains, up into its youngest and most rugged mountain ranges, and down into its deepest valleys. Looking down at our planet from space, we could at last see it whole.

We will never have that perspective, because we cannot see through even the clearest water for more than a few hundred feet. Insofar as we are interested in the shape of the ocean floor, the water of the ocean—teeming as it is with life, sustaining as it does all life on land—is merely an obstacle. Until this century the obstacle remained almost entirely impenetrable. If geologists have now penetrated it, and if they are now on the verge of constructing an image of the ocean floor that is for them the next best thing to a world without water, it is thanks above all to one fortunate fact: although water transmits light only poorly, it transmits sound exceedingly well.

Charles Bonneycastle's name does not appear in the *Dictionary of Scientific Biography,* or in the *Dictionary of American Biography,* although he was an American scientist, and a prescient one at that. Perhaps if the one experiment for which he is now remembered (and not by many) had been a success, things would have been different. In 1838 Bonneycastle, a professor at the University of Virginia, became the first man to attempt to determine the depth of the open ocean by echo sounding. He was about 75 years ahead of his time.

Bonneycastle had been inspired by an experiment done a few years earlier in Lake Geneva. On a clear night in November 1826, the Swiss mathematician Jean Daniel Colladon positioned an assistant in a boat on one side of the lake. A 140-pound church bell hung from the boat, a few feet below the surface of the water. As the assistant struck the bell with a hammer and simultaneously lit a flare, Colladon sat in a boat on the other side of the lake, eight and a half miles away. Seeing the flare, he listened for the sound of the bell with a 17-foot-long ear trumpet: a tin pipe, one end closed and submerged in the water, the other end open and conical and nestled over Colladon's ear. The sound of the bell came in clear but muted—like two knife blades banged together, Colladon said. From the delay between seeing the flare and hearing the bell, Colladon calculated the speed of sound underwater: it was 0.9 mile per second, more than four times the speed of sound in air. But that number, although it was the purpose of the experiment, was not the only thing that impressed Colladon. "The hearing of such a dry and short tone coming from many miles away," he wrote, "leaves an impression similar to that which one gets when one sees far-off objects appear clearly for the first time through a telescope."

Twelve years later, Bonneycastle was bobbing in the Gulf Stream, trying to see the ocean floor by means of sound. He was sitting in a boat detached from the U.S. Navy brig *Wash-*

ington, two days sail out of New York. He had a tin pipe similar to Colladon's pressed to his ear, but this time the cone was at the other end, underwater, and pointed toward the seafloor. A hundred and fifty yards away, the crew of the brig detonated a cast-iron petard at a depth of several fathoms. Bonneycastle heard the explosion—it sounded like knives struck together, he later said—and a third of a second later, he heard what might have been an echo from the bottom. A third of a second between explosion and echo translated, using Colladon's sound speed, into a depth of about 160 fathoms, or 960 feet. But when Bonneycastle sounded the depth with a line and lead, he found the depth was really closer to 540 fathoms.

The second sound Bonneycastle heard was probably a sort of aftershock, created by the implosion of a bubble of gas that expanded outward from the initial explosion. Why he did not hear an echo is not entirely clear. Perhaps the explosion was not powerful enough, or the noise of the sea surface was too great, or the surface area of his listening cone was too small to pick up the faint echo. A few years later Colladon did another experiment in Lake Geneva, this time with an 1,100-pound church bell, which he found he could hear from nearly 22 miles away. More interesting, a companion of Colladon's, sitting in a boat a couple of miles from the ringing church bell, appears to have heard an echo off the lake bank behind the bell. It ought to have been possible, even in the mid–nineteenth century, to have heard an echo off the ocean floor. But neither Bonneycastle nor Colladon seems to have pursued the matter. For the rest of the century, echo sounding was dead.

Sounding by the ancient method of line and sinker was very much alive, of course, and it was beginning to be pursued scientifically even as Bonneycastle was failing to start a revolution. From earliest times sailors had dropped lines over the side as their ship approached land, with a view toward not running aground. (The archetypical story is that of Saint Paul in the tempest, Acts 27:27–29: ". . . about midnight the shipmen deemed that they drew near to some country; / And sounded, and found it twenty fathoms: and when they had gone a little further, they sounded again, and found it fifteen fathoms. / Then fearing lest we should have fallen on rocks, they cast four anchors out of the stern, and wished for the day.") These efforts had nothing to do with science, and for obvious reasons they did not extend to the deep sea. It is said that in 1521, Ferdinand Magellan ran off 400 fathoms of sounding line in the Pacific out of pure curiosity and, failing to touch bottom, concluded somewhat illogically that he had found the deepest part of the ocean. But as an oceanographic milestone the story seems to be a myth; Magellan and his starving crew were approaching one of the northeastern Tuamotu Islands at the time and apparently were just looking in vain for an anchorage. Until the nineteenth century there was no will to explore the deep-ocean floor for its own sake, and thus there was no way.

The technical obstacles were twofold. To sound the bottom, you have to know when you have found the bottom, and in the open ocean the depths are so great that the weight of the line itself is enough to keep pulling it out even after the sinker is resting in the mud. That can fool you into believing the depth is greater than it really is. The second problem also leads to overestimated depths: ships tend to drift, particularly if there is a strong current, and so the sounding line is never quite vertical. In the 1830s and 1840s, as British and American sea captains began to adopt random deep-water sounding as an occasional pasttime, the public was regaled with a series of reports of incredible depths in the Atlantic— of 50,000 feet of line run out, say, with no bottom found. The reports were all quite wrong.

In the nineteenth century the only way to reduce the error that resulted from current drift was to try to steer the ship into the current or to drop the sounding lead from a longboat whose oarsmen kept it from drifting. And the way to know when you had hit bottom was to time intervals at which 100-fathom lengths of sounding line ran out and to watch for a slowdown indicating that the sinker had stopped sinking. Hanging a heavy weight from a thin line helped to make the change in speed recognizable. The prototypical sounding apparatus of the late nineteenth century was invented in 1853 by the American naval midshipman John Mercer Brooke: It used thin twine as a sounding line, and its sinker was a 63-pound cannonball. The ingenious thing about it was a system of hooks and lines that released the cannonball on impact. That allowed the line to be reeled back in—otherwise it would have snapped under the weight of the cannonball—and to bring with it a sample of the seafloor mud, pressed into a depression at the end of the iron bar from which the cannonball had hung. With a bit of mud in hand, you could be sure you had found bottom.

The U.S. Navy Depot of Charts and Instruments, under its director Matthew Fontaine Maury, used Brooke's sounding apparatus to complete the first organized survey of a large area of seafloor: the North Atlantic. From its ship the *Dolphin*, the depot made 200 soundings in the Atlantic between 52 degrees north and 10 degrees south. In 1854 Maury published the first chart of an entire ocean basin. It showed contours at 1,000, 2,000, 3,000, and 4,000 fathoms. It showed correctly that the bottom fell off rapidly at the edge of the continental shelf and that it rose again near the center of the ocean, to a broad elevation that Maury dubbed the Dolphin Rise. "The wonders of the sea are as marvelous as the glories of the heavens," Maury concluded, "and they proclaim, in songs divine, that they too are the work of holy fingers. . . . Could the waters of the Atlantic be drawn off so as to expose to view this great sea-gash which separates continents, and extends from the Arctic to the Antarctic, it would present a scene the most rugged, grand and imposing." Maury had a mystical bent.

As it turned out, he was right about the ruggedness of the seafloor, but it was just a lucky guess. Again, Maury was working with a mere 200 soundings. They were tens and even hundreds of miles apart, and each one—each depth and ounce or two of mud—had to be the synecdoche for tens of thousands of square miles of varied terrain. Nor were the soundings evenly spaced; a bunch were in the Caribbean, another bunch along the shipping routes to Europe. With 200 data points, Maury attempted to encapsulate an ocean larger by far than North America. Knowing the altitude of 200 points on North America, you might recognize that the Rockies are higher than the Great Plains, but you might also miss the Appalachians altogether. You certainly could not see ruggedness. You could only envision it, as Maury did.

He also envisioned, on the basis of 30 of his soundings, a "Telegraphic Plateau" between Newfoundland and Ireland. "The bottom of the sea between the two places," Maury said, "is a plateau which seems to have been placed there for the purpose of holding the wires of a submarine telegraph." In 1858 the Atlantic Telegraph Company, led by an American entrepreneur named Cyrus Field, laid the first transoceanic cable along Maury's plateau. It failed within a month. (Field received the last successful transmission, asking him to let the U.S. government know that the cable was ready for business, at a banquet held in New York to toast his achievement.) The cause of failure was probably the poor design of the cable, but it is also true that the Telegraphic Plateau did not exist. The first cable had been draped across mountains and canyons. In crossing the Atlantic they are unavoidable.

The extent of the mountain range that runs down the middle of the ocean—the Mid-Atlantic Ridge, it is now called—was discovered in the 1870s by the mother ship of oceanography, H.M.S. *Challenger*. The *Challenger* was a converted British warship, a 2,300-ton, 226-foot-long sail- and steam-powered corvette that had surrendered all but two of her guns to make way for naturalists and their gear. In December 1872 she embarked from Portsmouth, England, on the first global oceanographic expedition. In her long zig-zag stroll around the world ocean (68,930 miles in three and a half years works out to an average speed of two and a quarter miles per hour), she collected 13,000 different kinds of animals and plants and 1,441 water samples. She made hundreds of soundings and dredged hundreds of samples of seafloor mud and rock.

For the first time, she made the exploration of the deep ocean routine and even tedious. To measure the depth of the ocean at a single station and dredge a sample of the bottom was a long day's work, a day spent tossing on the waves and listening to the whine of the steam winch as it reeled in, first the sounding line, then the dredge. Moreover, there turned out to be a certain sameness to deep-sea mud. "At first, when the dredge came up, every man and boy in the ship who could possibly slip away, crowded round it, to see what had been fished up," one of the *Challenger's* naturalists wrote in his diary. "Gradually, as the novelty of the thing wore off, the crowd became smaller and smaller . . . at the critical moment, especially when this occurred in the middle of dinner-time, as it had an unfortunate propensity of doing. It is possible even for a naturalist to get weary even of deep-sea dredging."

But routine and tedious data gathering was what was needed to begin, as *Challenger* did, the systematic exploration of the ocean. Before the *Challenger* sailed, the ocean floor was essentially a blank slate—notwithstanding Maury's work and that of a few other expeditions of the mid–nineteenth century—and as such it was a perfect medium for all manner of speculation. What lay hidden beneath the Atlantic waves? Perhaps a lost continent, thought men as eminent as Thomas Huxley, defender of the theory of evolution, and Charles Lyell, pioneer of modern geology. Not only would a sunken continent comport with legend, but also it would help explain why fossils and rocks on one side of the Atlantic had close relatives on the other. As ancient continents sank to become oceans, ocean basins rose to become continents, and deep-sea mud, full of the chalky shells of microscopic organisms, became the white cliffs of Dover.

The belief in drowned continents and the "continuity of the chalk" was widely held, but far from universally. Other scientists clung with equal passion and equal ignorance to the view that continents and ocean basins could not possibly trade places, but were instead fixed and eternal, as ancient as Earth itself. The debate between these two opposing positions divided even the scientific staff of the *Challenger*. Although the *Challenger* expedition ought to have settled the issue, it did not: it was the beginning of real information about the ocean, but only the beginning.

In crossing the Atlantic five times, at latitudes ranging from the Azores in the north to Tristan da Cunha in the south, and the Indian and Pacific Oceans once, from west to east, the *Challenger* discovered the two most important geologic features of the planet. One was the first oceanic trench. On March 23, 1875, near Guam, at a place where her crew suspected nothing special, she ran out 4,475 fathoms of sounding line—more than 5 miles. Unlike the early and erroneous reports of great deeps in the Atlantic, and unlike the sounding of 4,655 fathoms reported the previous year by the American ship *Tuscarora* (which later proved to be accurate), the *Challenger's* claim was buttressed by a sample of the bottom clay. The Challenger Deep, as it is now called, is situated in the Mariana Trench, which is one of a series of deep trenches that rim the Pacific. Around 50 miles to the west of where the *Challenger* chanced to stop, the Challenger Deep plunges to 36,200 feet, or 6,033 fathoms, or 6.85 miles. It is the deepest spot in the ocean, deep enough to hide Everest easily, with room on top for a small Alp.

The other great geologic feature discovered by the *Challenger*—rediscovered and extended really—was the Mid-

Atlantic Ridge. On the home stretch in 1876, sailing north from Tristan da Cunha to Ascension Island, the ship found herself in consistently and surprisingly shallow water; she was sailing, in fact, along the crest of the ridge. The idea that Maury's Dolphin Rise might extend into the South Atlantic had suggested itself on the voyage out, on the crossing from the Cape Verde Islands to Brazil, when temperature measurements had revealed that the deep water near Africa was 2 degrees Fahrenheit warmer than the deep water near South America. It was logical to suppose that the two bodies of water were separated by some physical barrier. Now, as the sounding line repeatedly showed depths of less than 2,000 fathoms in the middle of the ocean, the idea took hold among the *Challenger's* scientists of a continuous submarine mountain range running from Iceland to Tristan and paralleling the coasts.

When news of the discovery was received in London, it was promptly hailed as the discovery of Atlantis. There was a similar reaction to other midocean ridges the *Challenger* brought to light; the Carlsberg Ridge between India and Africa, for instance, was greeted as Lemuria, a drowned continent that once united the similar faunas of India and Madagascar (although lemurs themselves live only on Madagascar). The *Challenger* collected plenty of evidence to puncture these flights of fancy. None of her samples of deep-sea mud contained organisms resembling the fossils in England's chalk cliffs (which are now known to have been deposited in shallow water). With the exception of a few rocks that had apparently been carried out to sea and dropped by glaciers, none of the rocks she dredged from the deep sea were continental rocks— granites, say. In short, there was no evidence that the undersea mountain ranges had once been continents or that the continents had once been deep-sea floor. But the idea lived on nevertheless. It survived in part because it fulfilled a need: the need to explain why fossils could be so similar when an ocean lay between them. It also survived because the few hundred stations occupied by the *Challenger,* in 142 million square miles of ocean, were not enough to dislodge strongly held beliefs.

At the turn of the century, it was still possible to believe, and many geologists did, that Earth was shrinking as it lost its primordial heat; that as it contracted, blocks of crust subsided to become ocean basins, while others were squeezed upward to form mountain ranges and whole continents; and that low points became high and high points became low as the cooling continued and the erosive forces of wind and rain did their steady work on the land. On the other hand, it was also possible to believe, as the *Challenger's* scientists had come to believe, that the present ocean basins were permanent features of Earth's surface. And finally, it was possible to believe, as a German meteorologist and visionary named Alfred Wegener believed, that continents drifted horizontally across the face of the Earth, plowing through the ocean basins—which Wegener claimed were basically flat and featureless—as ships plow through water.

It was possible to believe many things about the ocean in the early years of this century, because so few things were known. In 1904, eight years before Wegener advanced the theory that later proved so prescient, the newly formed International Hydrographic Bureau had prepared the first standardized bathymetric chart of the world ocean, collecting all the available soundings. There were 18,400 of them. That was not enough to understand the ocean floor, and in fact geologists did not begin to understand the overall architecture of the ocean floor—or the planet as a whole—until half a century later. Until then, they really did not know what they were talking about.

If a simple desire for understanding had been the only motive force, they might never have found out. Just as the goal of laying a transatlantic telegraph wire had justified and paid for some of the earliest deep-ocean soundings, so too did eminently practical concerns, rather than a lust for knowledge, drive the realization in this century of Bonneycastle's dream of echo sounding the ocean floor. Those concerns may be summed up in two words: *Titanic* and *Lusitania*.

After the *Titanic* slammed into her iceberg in 1912, at a cost of 1,513 lives, inventors of devices for detecting underwater hazards began crawling out of the woodwork. One of them was Reginald A. Fessenden of the Submarine Signal Company in Boston, which for a decade or so had been attaching underwater bells to lightships and equipping ships with microphones to hear the bells. Fessenden invented a sound source that was much louder than a bell—loud enough for ships to use to echo-locate an iceberg, say, that was not equipped with a warning bell. In the Fessenden oscillator, an alternating current in a primary coil induced a secondary current in a copper cylinder that caused the cylinder to vibrate back and forth, warping a large metal membrane and emitting a loud metallic noise. The noise was so loud that on an early field test of the device, on a Coast Guard cutter off the Grand Banks, officers sitting in the cutter's mess could hear, without benefit of microphones, the sound echoing off an iceberg two miles away. They could also hear an echo off the ocean floor, one mile down.

Fessenden was primarily interested in detecting icebergs and German submarines—like the one that sank the *Lusitania*—not the ocean floor, and although he did go on to build an echo-sounding device, it was not a successful one. His oscillator, though, became the sound source of the first practical deep-ocean echo sounder, which was invented by a U.S. Navy physicist named Harvey C. Hayes.

In principle it is easy to echo-sound the depth of the ocean: you just transmit a sound downward and time the interval until you hear the echo. But in practice it was difficult to measure the interval accurately enough; an error of half a second translates into a depth error of more than a thousand feet. The Hayes Sonic Depth Finder overcame this problem. It transmitted pulses of sound at an interval that could be varied by the operator. The operator wore earphones, one ear connected to the transmitter, the other to the echo receiver. He adjusted the pulse interval until the sound was centered, that is, until the echo from some earlier pulse was arriving at the exact instant that another pulse was being transmitted. The number of sound pulses traveling in the water at that instant was a whole number. The distance to the seafloor and back was equal to that unknown number multiplied by the known distance that sound could travel in the time between pulses. To find out the number, and thus the depth, the operator varied the pulse interval slowly until he again had a transmitted pulse and a received echo arriving simultaneously at his ears. Now he had two equations for the depth: the first representing it in terms of the original unknown number of pulses in the water, the second representing it in terms of that number plus one. Two equations and two unknowns—that is junior-high algebra, and the operator did not even have to bother with algebra, because Hayes prepared tables that spared him the trouble. A single deep-ocean sounding with line and sinker had taken the better part of a day; with the Hayes Sonic Depth Finder a sounding could be executed in a minute.

That meant it was now possible to prepare quasi-continuous profiles of the ocean depth along the track of a ship. Hayes himself made the first such profile across an entire ocean, sailing from Newport to Gibraltar aboard a destroyer, the U.S.S. *Stewart*, between June 22 and June 29, 1922. During that week he made 900 deep-sea soundings—three times the number *Challenger* had managed in a three-and-a-half-year cruise. Hayes's invention had opened a door into the unknown realm that is most of our planet.

The first oceanographic vessel to drive through that door

was the German ship *Meteor*, which crisscrossed the Atlantic from 1925 to 1927. Her main mission was to study ocean chemistry; among other things, the scientists aboard her were supposed to determine whether enough gold could be extracted from seawater to pay off Germany's crushing debts from World War I. (It could not.) But the *Meteor* took echo soundings every 5 to 20 miles, and so produced relatively detailed profiles of the ocean bottom under her track. The profiles confirmed for the first time that the Mid-Atlantic Ridge was not a smooth plateau but a rugged mountain range—as Maury had envisioned it, 70 years earlier, on the basis of no evidence at all.

After World War II, the Woods Hole Oceanographic Institution's *Atlantis* undertook a series of expeditions to the Mid-Atlantic Ridge. That ship was equipped with a "fathometer" that could make continuous bottom profiles, as long as the power supply was willing. The power supply was not always willing: Whenever someone opened the ship's refrigerator, the power to the echo sounder was cut off, and it recorded a bottomless depth. But that was a small inconvenience. The fathometer and the other echo sounders of the day were far from perfect, and they were terribly primitive compared with the sonar mapping devices available now. But the information they provided was good enough to produce one of the most important geological discoveries of the twentieth century.

———

Marie Tharp receives visitors these days in a room that is about ten feet away from the Hudson River at its widest and most oceanic point—the Tappan Zee—and that is part living room, part bedroom, and part explorer's den. The room is decorated with globes large and small, with African masks and oil seascapes, with stacks of books and papers and assorted mementos, and, most noticeably, with photographs of Bruce Heezen. Next to one of the photographs is a plaque commemorating the NR-1, a Navy research submarine, in which Heezen died of a heart attack in 1977. The sub was headed for the floor of the Atlantic Ocean at the time, which means Heezen died at home. He was a marine geologist. He was also Tharp's companion and her boss. He is still a constant presence in her life and conversation—conversation that meanders somewhat vaguely between moments of precise recollection. Tharp recalls clearly the events of 1952, when she was working for Heezen as a research assistant at the Lamont Geological Observatory, 10 miles down the road on the Palisades. It was in that year that she discovered the rift that runs down the middle of the Mid-Atlantic Ridge. She discovered it, and Heezen understood what it meant.

Son of an Iowa turkey farmer, Heezen was plucked out of the University of Iowa in 1947 by Maurice Ewing, founder of the Lamont lab. As Tharp tells the story, Ewing was a visiting lecturer at Iowa, was introduced to Heezen, then a junior, and promptly said: "Young man, would you like to go on an expedition to the Mid-Atlantic Ridge? There are some mountains out there and we don't know which way they run." The Atlantic was wide-open spaces to an Iowa boy, intellectually as well as geographically. Within a year Heezen was studying the Mid-Atlantic Ridge, and within five years he was mapping it. (For some 20 years he ate no turkey.)

It was Tharp, his research assistant, who actually drew the map. And what she drew at first, sitting on the second floor of what had been the Lamont family manor (Heezen's private study was a bathroom that overlooked the Hudson) was not a map at all but a series of six detailed depth profiles. Each profile represented the shape of the ocean floor along a single line across the entire North Atlantic, and each had to be laboriously pieced together from echo soundings collected by the *Atlantis*, the *Meteor*, and even the *Stewart* on its pioneering voyage three decades earlier. Tharp's profiles all agreed on the general features of the Atlantic basin. They showed its basic

east-west symmetry: the steep descent from the American continental shelf down onto a broad abyssal plain, punctuated by
isolated mountains such as Bermuda; the very gradual rise up
the flank of the Mid-Atlantic Ridge, which occupies the center
third of the basin; and finally the descent down the eastern
flank, into an abyss that abuts the continental shelves of
Europe and Africa. The profiles agreed on these general features, but none of their details matched up—with one exception. On the three northernmost profiles in particular, Tharp
noticed a V-shaped notch in the very crest of the Mid-Atlantic
Ridge.

The notch does not leap out at someone looking at the profiles today, and it did not leap out at Heezen then. Later he
was to recall his reaction when Tharp pointed it out to him:
he dismissed it as "girl talk." "It cannot be," Tharp remembers
him saying. "It looks too much like continental drift."

In the four decades since Alfred Wegener had noted the
remarkable fit between the coastlines of the Americas on the
one hand and Europe and Africa on the other and had suggested that all those continents had once been joined together,
the expert reaction to his theory of rifted and drifting continents had progressed from skepticism to ridicule. This was
particularly true on Heezen's side of the Atlantic. To espouse
the drift theory as a geologist was to risk one's career. Although
Wegener himself had known little of the Mid-Atlantic Ridge—
he thought it might be a pile of rubble left behind by the
separating continents—the notch that Tharp was now proposing to locate down the center of the Ridge looked disturbingly
like a Wegenerian rift. Heezen was not ready to see it.

Two things finally opened his eyes. One was a study of the
rift valley of East Africa—the long trough that snakes through
the continent from the Gulf of Aden down to Lake Malawi.
Geologists agreed by then that the East African valley was a
rift—a crack in Earth's crust where it was being pulled apart.
When Tharp plotted topographic profiles across the African

rift, they looked much like her profiles across the Mid-Atlantic Ridge.

The other eye-opener for Heezen was a second type of sound record of the ocean floor. This was a record not of the sounds bounced off it by geologists but of the sounds it made itself: the low inaudible rumble of earthquakes. Seismic sound waves travel long distances through solid rock, and so an earthquake on the ocean floor could be detected by seismometers on land. In the 1940s the seismologists Charles Richter (of Richter scale fame) and Beno Gutenberg had noted that the Mid-Atlantic Ridge was a belt of earthquake epicenters. When Heezen had a second draftsman, working at the table next to Tharp's, plot the earthquake data at the same scale at which she was plotting the seafloor topography, an important connection emerged. Not only did the epicenters fall within the 1,000-mile-wide Mid-Atlantic Ridge, but also virtually all of them fell, within the measurements' margin of error, inside Tharp's V-shaped notch, at the very center of the ridge.

The conclusion now seemed clear. That notch was a rift valley just like the rift valley in East Africa. It was a place where the seafloor was cracking apart, generating earthquakes in the process. What is more, seafloor cracking was not confined to the mid-Atlantic. That was only where the most detailed echo-sounding records were to be had. But the record of earthquake epicenters was much more extensive, and it revealed belts of earthquakes winding throughout the world ocean, in regions that had never been sounded. Heezen made a bold but logical leap: if the earthquakes in the Atlantic occurred in a rifted ridge, then the ones in other oceans probably did as well.

Heezen and Ewing thereupon proposed that a continuous ridge, with a rift valley running down its center, extended around the world. It wound around the Cape of Good Hope, through the Indian Ocean south of Australia, across the South Pacific, and up the west coast of the Americas. In the Indian

Ocean a spur called the Carlsberg Ridge—which had been sounded by the *Challenger* and by a Danish expedition—came ashore in the Gulf of Aden, branching to the north into the Red Sea and to the south into the African rift. In Iceland, which was also known to have a prominent rift valley, the Mid-Atlantic Ridge came ashore. It then continued north through the Arctic Ocean, past the North Pole, before vanishing somewhere in eastern Siberia. All around the world the rifted ridge continued, like the seams of a baseball, or as Heezen put it, like a wound that never heals.

For this valley—this canyon, really: the drop from the mid-Atlantic peaks to the bottom of the rift can be well over a mile, deeper than the Grand Canyon—was no peaceful river cut. And the ridge itself was certainly not a sunken continent, nor even a pile of rubble swept off the continents. Oceanographers dredging the ridge after World War II had been no more able than H.M.S. *Challenger's* crew to find continental rocks on it; what they found was volcanic basalt. The diagnosis was confirmed by a third kind of seafloor sound record, a technology pioneered by Ewing: seismic shooting. Ewing used sticks of dynamite to create sound so powerful that, instead of bouncing off the seafloor like a sonar ping, it would penetrate the floor and bounce off buried layers of crust. Using this method, he and other oceanographers were able, beginning in the early 1950s, to measure the thickness of the ocean crust and of the layer of sediment that covered it. The seismic shooters could also determine the speed of sound in the various layers, which gave them an idea of what the layers were made of. At the Mid-Atlantic Ridge, there was almost no sediment covering the crust, and the crust was made of basalt and other volcanic rocks.

Hence the "wound that never heals": In Heezen's view the mid-Atlantic rift was a volcanic fissure, ever widening, pushing the rest of the Atlantic before it, and ever being filled from below by lava welling up from Earth's hot, rocky mantle. By the time Heezen suggested this idea, in the late 1950s, the taboo against continental drift was beginning to lose its force. It had been undermined in particular by studies of the magnetism of ancient rocks on land. When magma cools and solidifies or when sediment is deposited, magnetic crystals inside the nascent rock align themselves with Earth's magnetic field and are frozen in that position. In the 1940s and 1950s, British geologists discovered that these fossil compass needles did not necessarily point anywhere near today's magnetic North Pole. Either magnetic North had moved by a substantial amount or the continents carrying the rocks had—and the latter theory proved the only way to make consistent sense of data from different parts of the world. Alfred Wegener, who had frozen to death on the Greenland ice sheet in 1930, thus began to acquire a posthumous following. And this time, thanks to Heezen and Tharp's discovery, one no longer had to imagine the continents cutting their way through solid ocean floor like icebreakers. Instead, the continents were moving apart because new ocean crust was constantly being created between them, at midocean rifts.

There then arose an obvious question: What was happening to the old ocean crust? Where was it going? Heezen had an answer: the ocean crust was not going anywhere; Earth was expanding. At one point, Heezen argued, granite continents had covered the entire globe in an unbroken shell: but as the planet had ballooned outward, the shell had cracked open and oceans had appeared in the gaps. This idea was even bolder than the idea of a world-girdling rift. But here Heezen's boldness failed him. In 1957, when he went to Princeton to give a talk on midocean rifts, the chairman of the geology department came up to him afterward and said, as Tharp recalls it, "Young man, you have shaken the foundations of geology." Heezen shook the old foundations, alright, but then he just kept on shaking; it was that chairman, Harry Hess, who built the new ones.

Hess's story has often been told, and a great deal has been made of his wartime experience as a naval commander in the Pacific, when he kept the ship's sonar running at all hours in the service of oceanography. He will be remembered, though, not for his tenacity as a data gatherer but for his genius as a synthesizer—as a writer, to use his word, of "geopoetry." Hess put everything together, at a time when he had long since stopped going to sea.

By 1960, when Hess wrote his seminal paper (later published as "History of Ocean Basins") there were a number of disparate observations that needed putting together. There was Heezen and Tharp and Ewing's discovery of rifted midocean ridges. There were measurements showing that the ridges were made of basalt, had a great deal of heat escaping from them, and, at least in the Atlantic and Indian Oceans, ran exactly down the center of the ocean basins. There was the suggestion, made nearly three decades earlier by the British geophysicist Arthur Holmes, that all the heat generated by radioactive elements in Earth's interior ought to roil its rocky mantle like a pot of boiling water. A midocean ridge, Holmes had speculated, might be a place where one of those convective currents rose to the surface.

Finally, there were the observations of a Dutch geophysicist named Felix Andries Vening Meinesz, a man who was too tall to stand upright in a submarine, but who nevertheless logged 125,000 miles in submarines between 1923 and 1938. He did it to make precise measurements of how Earth's gravity field varied from place to place in the ocean. As a graduate student in 1932, Hess had accompanied Vening Meinesz on a cruise to the Puerto Rico Trench, the deepest spot in the Atlantic, where the ocean floor plunges to a depth of 28,232 feet. Vening Meinesz had determined that gravity there and at other oceanic trenches was anomalously low—perhaps, he suggested, because a trench was a place where lightweight, low-gravity crust was being forced downward and was displacing denser, higher-gravity mantle. This observation was to prove especially significant to Hess in 1960. It gave him something Heezen had lacked: a place to put the crust that was being made at midocean ridges, without resorting to the wild idea that the planet was expanding. The one great flaw of Heezen's theory was that it only took account of one of the two most important features of the ocean floor; it ignored the deep trenches.

In hindsight what Hess made of all these observations seems almost obvious; indeed, a Navy geophysicist named Robert Dietz invented nearly the same theory simultaneously and independently, and the basic idea had been foreseen by Arthur Holmes. But it is often thus with intellectual break-throughs. Hess's theory, which is now accepted as fact, is called seafloor spreading. It is this simple: Earth's mantle is indeed in convective motion, as Holmes had suggested; it is solid rock but so hot that it slowly creeps. Like a pan of boiling water, it rolls. Hot mantle rock rises to the surface at a midocean ridge—more precisely, in the rift valley that cuts the crest of the ridge—and forms new seafloor. The seafloor spreads away from the ridge on both sides and cools as it does, just as water in a pan spreads away after boiling to the top. Where the seafloor runs into a continent, or simply gets old, cold, and dense enough, it sinks down into the mantle. The result is an ocean trench, with the low gravity that Vening Meinesz observed.

In other words, the amount of seafloor is not constantly increasing, as Heezen thought. Instead old seafloor is constantly being recycled back into the mantle at trenches. And what of the continents? They drift, as Wegener had said, but they are not in charge of the drifting. "The continents do not plow through oceanic crust impelled by unknown forces,"

Hess wrote, "rather they ride passively on mantle material as it comes to the surface at the crest of the ridge and then moves laterally away from it."

Hess's concept of seafloor spreading was the beginning of a revolution in geology, as complete a revolution as any scientific discipline has ever known. Two things had to occur to finish what Hess started. First it had to be shown that seafloor spreading actually happened. The proof came in 1966, from observations of magnetic trails that the spreading seafloor leaves in its wake. Basalt solidifying from molten rock at a midocean ridge acquires the prevailing magnetic field of the day and keeps it as it spreads away from the ridge. But from time to time, over millions of years, the direction of the Earth's magnetic field flip-flops; the poles change places. Those flip-flops turned out to be recorded in the magnetism of the seafloor, in a pattern of stripes that paralleled the mid-ocean ridge on either side. Each stripe was mirrored by an opposite number, with the same magnetic polarity and at the same distance from the ridge. Such a pattern could only have been made by seafloor spreading: the mirror-image stripes were ribbons of rock that had crystallized at the ridge at the same time and then moved away in opposite directions.

The second phase of the revolution involved figuring out how seafloor spreading worked on a spheroidal planet, with its seemingly disorderly array of ridges and trenches. It was a problem in geometry, and it was solved in 1967 by a Princeton geophysicist named Jason Morgan (and, independently, by a geophysicist from Cambridge University named Dan McKenzie). Just as Hess had seen farther than Heezen by considering trenches as well as ridges, Morgan finished what Hess had started by incorporating a third type of seafloor feature: the oceanic fracture zones.

Those deep troughs cut through the continuous line of the midocean ridge at frequent intervals and at right angles. They shift the crest of the ridge to one side or the other by as much as hundreds of miles. Some of them, in the equatorial Atlantic for instance, continue on past the ridge and cross an entire ocean basin. The fracture zones are massive disruptions in the shape of the ocean floor, and they were discovered by echo sounding in the early 1950s. But it took Hess's theory of sea-floor spreading for their significance to be appreciated—that and a flash of insight in the brain of a Canadian geologist named J. Tuzo Wilson. Wilson saw that the part of a fracture zone situated between two offset segments of a ridge was a new kind of fault, which he called a transform fault. Whereas at ridges new crust is created and at trenches old crust is destroyed, at transform faults blocks of crust simply slide past each another. The fracture zones away from the ridge are just the relics of those transforms—frozen streamlines that reveal the relative motion of the blocks.

Three types of fault and three types of motion were enough for Morgan. The fracture zones crossing any given midocean ridge, he recognized, are segments of concentric circles; this means the crustal motion they trace is a rotation about a common center. Taking Heezen and Tharp's map of the South Atlantic, he showed that the fracture zones cutting across the Mid-Atlantic Ridge could all be approximated by concentric arcs centered on a "pole of rotation" at 65 degrees north latitude and 35 degrees west longitude. On the other hand, the fracture zones in the Pacific, off the coast of California and Central America, had a common center at 79 degrees north and 111 degrees east. Morgan realized that the way to accommodate all the different rotations—all the different ridges, trenches, and fracture zones—in a single consistent scheme was to imagine the whole of Earth's crust as being broken into rigid blocks. The blocks slid separately over the underlying mantle, at once jostling and being constrained by their neighbors, in an ever-shifting tilework with never a gap.

On April 17, 1967, at a meeting in Washington of the American Geophysical Union, Morgan presented this theory for the first time. It was a theory he had been working on excitedly for just a few weeks. Now he was presenting it at a sparsely attended session—most of his audience had left for lunch—at which he was supposed to be talking about something else altogether. He plunged right in:

> Consider the earth's surface to be made of a number of rigid crustal blocks, and each block is bounded by rises (where new surface is formed), trenches or folding mountains (where surface is being destroyed), and great faults. Assume that there is no stretching, folding, or any distortion within a given block. Then, on a spherical surface, the motion of one block (over the mantle) relative to another block may be described by a rotation of one relative to the other.

Replace the word "block" with "plate" and read "transform fault" or "fracture zone" for Morgan's "great fault," and you have the theory we now call plate tectonics, in a nutshell, from the man who invented it. The response that day in Washington, however, was indifferent. Even an audience of seafloor-spreading devotees could not quite see why they should "consider the earth's surface to be made up of a number of rigid crustal blocks" merely on the say-so of a young and little-known theorist from Princeton. Morgan was ahead of his time—but only just. Within a year, after he and McKenzie had both published papers on the subject, the walls of the old "fixist" Earth view had begun to crumble. Almost overnight they turned to rubble.

With plate tectonics, the revolution launched by Heezen and Hess (and a few others, of course) was complete. A whole new way of looking at the planet, the first real way of looking at the whole planet, was now in place, and it provided a globally coherent explanation for the grandest features of Earth's surface. Geology, it was now clear, happened first and foremost at plate boundaries. A volcano erupting in the Philippines or Japan or the Aleutians, for instance, could trace its roots to the Philippine or Japan or Aleutian trench, where the Pacific plate plunged under a neighboring plate, releasing the heat and molten rock that pushed up the volcano; and that same volcano had roots even farther afield, at the East Pacific Rise, where the Pacific ocean floor was born from upwelling magma. The earthquakes that rocked California with grim frequency could be traced to the same East Pacific Rise but also to the Mid-Atlantic Ridge, for the Atlantic was growing at the expense of the Pacific, and the westward drift of North America had caused it to bury one of the biggest fracture zones of the eastern Pacific—a fault now called the San Andreas. In the Red Sea and the East African rift valleys, a new ocean was being born from a sundered continent. In Asia, an old ocean, or part of one, had already been destroyed, when India rammed into the continent with enough speed to push up the Himalayas.

No major geologic feature could escape being reinterpreted in plate tectonic terms. But it was our understanding of the ocean that changed most dramatically between 1952 and 1967. The change had begun with Heezen and Tharp's discovery, and it had gathered steam with the discovery by Ewing and others that the sediment cover on the ocean floor was extremely thin. On average there was less than a mile of it. That meant that even if the sediment had accumulated at a rate much lower than the rate at which dust collects on your furniture, the ocean floor could not have been collecting sediment for billions of years; in fact, as Hess calculated in his famous paper, it could not be more than about 260 million years old in its oldest regions. Thus the ocean was not merely the ancient trash dump of the continents, a catchment basin for continental debris and the corpses of marine organisms (although it was that too). The ocean was young.

And it was active. With seafloor spreading and plate tectonics, the notion that geology really happened on land—a notion that even Alfred Wegener had shared—was finally exploded. One reason Wegener's theory had never gained wide acceptance was that he had no plausible mechanism for pushing the continents around the globe. The problem, in retrospect, was his and everyone else's ignorance of the ocean. It was not so much that the continents were drifting; it was that the seafloor was spreading. The continents were just along for the ride. The ocean floor was where the action was, and in the 1950s and 1960s, our towering land-centered ignorance of it began at last to lift.

CHAPTER 3

To Map Is
to Know

Whan Bruce Heezen sailed out to the Mid-Atlantic Ridge for the last time in 1977, he carried with him the chromalin proofs for a new map he and Tharp had been working on for some years: a panorama of the entire ocean floor. It was a physical map, like the ones you see of continents in atlases, which show you roughly what the land would look like from space on a cloudless day. Even today such maps generally color the ocean a uniform bright blue, omitting any depiction of the terrain underneath the water. The point of Heezen and Tharp's map, the first of its kind, was to fill in the blank blue—to show what the ocean might look like on a waterless day.

Before Heezen and Tharp came along, seafloor maps, beginning with Maury's in the nineteenth century, had been contour maps. They were flat abstractions, good for showing sailors where they might run aground but wholly inadequate for showing what the ocean floor looked like. When Heezen and Tharp started their work in the early 1950s, there were two reasons to choose a different approach. The first had to do with the Cold War. The U.S. Navy, planning for submarine warfare, had begun to classify contour maps of the ocean floor. Even if they had wanted to, Heezen and Tharp would not have been allowed to publish their data as contour maps. And one day in 1952 they decided they did not want to anyway. On that day Heezen sat down and, in about an hour, sketched a rough diagram of what the western North Atlantic really looked like, without water and from an oblique perspective, as best he could render it with his limited skills as a draftsman. He handed it to Tharp and asked her to do a better job.

Thus began a 25-year effort that gave the public their first worthwhile mental image of the two-thirds of Earth they could not see. Tharp and Heezen started in the early fifties with the North Atlantic, with the map that first showed the rift valley, and proceeded to the South Atlantic and thence to the Indian and Pacific Oceans. Beginning in 1967, the year plate tectonics exploded on the geologic scene, the seafloor maps were published in *National Geographic*—a perfect outlet for reaching the widest possible audience and for capturing the imagination of a generation of potential young oceanographers. The maps were certainly pretty enough. They were painted in glorious color from Tharp's line drawings by Heinrich Berann, an Austrian artist who had made a profession of painting Alpine panoramas for ski resorts and who knew how to paint a mountain.

Besides being attractive, the maps were also accurate in their depiction of the major provinces of the ocean floor: the ridges and trenches, the continental shelves and abyssal plains. They showed vividly, in a way that contour maps never could, how one province merged into another—for instance, how sediments from the continental shelves flowed down the continental slopes and then fanned out onto the abyssal plains.

Heezen and Ewing had helped figure out this process in the late 1940s when, on a hunch, they had examined the records of a curious seafloor incident. In 1929, a magnitude 7.2 earthquake on the Grand Banks, south of Newfoundland, had led to the rupture of a dozen or more undersea telegraph cables. But the cables had not been broken by the quake itself. They had broken hours later and as far as 300 miles from the epicenter, starting at the top of the continental slope and then proceeding downslope, snapping one after another like a string of firecrackers, all the way down to the Sohm Abyssal Plain.

To Heezen and Ewing, this was the unmistakable fingerprint of an earthquake-triggered avalanche—or, to use the technical term, of a "turbidity current" of water that was heavy with suspended sediment and was therefore inclined to cascade into the abyss. The speed of the Grand Banks turbidity current is now reckoned to have been as much as 45 miles per hour. Its range was far greater than any avalanche on land.

Turbidity currents are not always so dramatic, but their effects can be seen all over the Heezen-Tharp panorama. The currents are like underwater rivers, albeit intermittent and catastrophic ones, and like rivers they cut canyons into the margins of the continents, and great meandering channels down the continental slopes and out into the abyss. Sometimes the underwater currents are direct continuations of rivers on land, which are rich sources of sediment; the Hudson River canyon, for instance, carries New York State dirt more than 200 miles into the Atlantic. Sometimes the underwater rivers have no counterparts on land; the Monterey Canyon, which channels sediment to a depth of more than 15,000 feet and to a distance of some 300 miles from the California coast, is clearly not the handiwork of the sleepy Salinas River.

Whatever their connection to the land, though, turbidity currents have an extraordinary impact on the abyss. The most spectacular example on the Heezen-Tharp panorama are the vast fans of sediment that flank the coasts of India. Even as the Himalayas continue to be pushed up by the collision of subcontinent and continent, the peaks are being dismantled, grain by grain, by wind and rain and snow. These sediments are being carried into the Arabian Sea by the Indus River and into the Bay of Bengal by the Brahmaputra and the Ganges. There turbidity currents take over and spread the sediments far and wide. The Indus Fan covers all the floor of the Arabian Sea as far south as the tip of India. The Bengal Fan extends even farther, past Sri Lanka and out onto the Ceylon Abyssal Plain—a distance of nearly 2,000 miles. The Bengal Fan is about 600 miles wide, and in some places the sediments are more than 12 miles thick.

It is turbidity currents such as these, along with the slow rain of dead organisms from the surface of the sea, that conspire to make the abyssal plains the flattest landscape on Earth. Typically the slope on the plains is no more than 1 in 1,000; this means they rise or fall less than 2 yards in every mile. But without the sediments they would not be plains at all. The sediments cover the rough terrain that emerges from the midocean ridges, filling in the valleys and the gullies, burying all under a smooth and general blanket. As one ascends toward the midocean ridges, though, the roughness begins to show through, in reality as on the Heezen-Tharp map.

That ascent is a passage in time as well as space. The ocean floor far from a midocean ridge, along the edges of continents, is old ocean floor: it was created by magma welling up at the ridge millions of years ago and has since spread away from the ridge. As the rock spreads, it cools; as it cools, it contracts; and as it contracts, it subsides. Hence the great depth of the abyssal plains and their comparatively thick sediment cover: they are the oldest, coldest part of the seafloor, and they have had plenty of time to subside and be buried by sediment. Hence too the soaring heights of some of the midocean ridges. Unlike mountain ranges on land, they have not been pushed up above their surroundings: their surroundings have fallen away from them.

The ridges stand as much as 15,000 feet above the abyssal plains, but the ascent is gentle; halfway up the flank of the Mid-Atlantic Ridge, for instance, the slope is roughly the same as the slope of the Great Plains as they rise toward the Rockies. Were one to ascend this barely perceptible incline—something no human being has yet done, for there is no submarine that can dive that deep for long enough—one would begin to encounter the ridge's handiwork poking through the detritus of the ages. One would enter a landscape of abyssal hills, typically a few hundred feet high or so. It is a landscape that occupies vast areas of the ocean. Half the floor of the Atlantic is dotted with abyssal hills; in the Pacific the figure is closer to 80 percent, because so much of that huge ocean is far from continents and thus beyond the reach of turbidity currents. The Heezen-Tharp map shows the abyssal hills, but not very accurately: it depicts them as circular structures, whereas it is now known that they tend to be elongated sausages that parallel the midocean ridges.

The ridges themselves are the most prominent feature on the map. They form a single continuous ridge that snakes around the planet—submerged under California, interrupted and offset at regular intervals by long transform faults, but otherwise unbroken. Its position is everywhere highlighted by a dark central line, representing the shadowed depths of the rift valley. On the Heezen-Tharp map the rifted ridge looks essentially the same in the Atlantic as in the Pacific. That is another of the map's many limitations.

"I don't want to be critical of Marie," says Bill Ryan, who did his Ph.D. under Heezen and who is now one of the senior marine geologists at what is now called the Lamont-Doherty Earth Observatory. "What she did was wonderful in her day. She had an intuition of how the Earth worked, working with Bruce Heezen—a seafloor-spreading intuition, so they got a lot of it right. But it was an intuition. The maps are correct in the gross nature of where the ridges are, but her fabrics are wrong in lots of places. Her textures are wrong. She has at the most maybe a dozen different classes of landscapes, and each one is strongly stylized."

The most obvious way in which the various Heezen-Tharp maps differ from reality is one that geologists like Ryan accept as inevitable and do not even worry about, but one that must have produced a serious distortion in the general public's mental image of the seafloor. The maps exaggerate the vertical dimension of every seafloor feature by a factor of around 20. They have to exaggerate, because otherwise the ocean floor would look almost as featureless as people once thought it was. A 5-mile-high mountain depicted at its true scale would disappear on a map that spans 25,000 miles (Earth's circumference at the equator). The mountain's height must be exaggerated if it is to be visible at all.

But with the exaggeration comes a serious loss of realism. Stretching the vertical 20 times while leaving the horizontal untouched means that a three-story house become a skyscraper, and a needle-thin skyscraper at that. It means that undersea volcanoes become towering spires that would collapse under their own weight if they really existed. It means that the edge of the continental shelf off the Atlantic coast of North America becomes a sheer cliff, whereas in reality the slope down into the abyss averages about 1 in 14, something even an amateur could tackle on a bicycle, something trucks are not even asked to shift into low gear for. Lastly, the vertical exaggeration on the Heezen-Tharp maps means that the great trenches that ring the Pacific look like yawning chasms, whereas in reality you could probably drive your jeep into Challenger Deep if the water were drained and you took care to avoid the occasional precipitous ledge.

Vertical exaggeration is still present on seafloor maps today, and it is relatively easy to take into account if one is aware of it. That was not true of the more fundamental flaw of the Heezen-Tharp maps, a flaw that was also not the fault of Heezen or Tharp. To say that they relied heavily on intuition in sketching the seafloor is to engage in euphemism: they made most of it up. They had vastly more sounding data to work with than Matthew Fontaine Maury or any other cartographer, but given the size of the ocean, the amount of data they had was still ridiculously small. In drawing the Indian Ocean, for instance, Tharp confronted a stretch of midocean ridge more than 4,000 miles long between Antarctica and Australia that had only one line of soundings running across it. ("This section of ridge," she later wrote, "was of necessity sketched in a very stylized manner.") By comparison, the South Atlantic was relatively data-rich. Tharp had 30 ridge-crossing sounding lines to work with there, the ones collected by the *Meteor* in the 1920s—but those 30 lines of data were spread again over more than 4,000 miles of ridge. Even under favorable circumstances, in other words, the lines of reality were on average about 130 miles apart.

Tharp would plot these sounding lines as profiles, then convert them to oblique "aerial" views. In that simple act of expanding a two-dimensional cross section to a three-dimensional stripe of seafloor seen in perspective, there was already considerable invention. But even more invention was required to connect the stripes across hundreds of miles of unsounded seafloor. "How did I get from here to there?" Tharp recalls. "I filled it in!"

As a result, you cannot use a Heezen-Tharp map to lead you to a particular undersea mountain, because the mountain might not be there. The whole mountain range might be in the right location—might—but certainly not the individual peak. (As recently as 1984, oceanographers who went looking for the rift valley in the South Atlantic, using the Heezen-Tharp map as their guide, found that the valley was 150 miles away from where it was supposed to be.) Searching in vain for the mountains Tharp had drawn, you would run into ones that were not on her maps at all. They really were not maps so much as portraits of the ocean floor, and very successful ones. But it is unlikely that many of the millions of people who have perused the Heezen-Tharp maps and their intellectual derivatives over the years have understood the extent to which the image deviated from reality. Very few people appreciate just how little is known about the ocean floor, even today.

For decades now oceanographers have been fond of pointing out that we know more about the surface of the moon than about the seafloor; anyone with a backyard telescope can get a clearer view of lunar geology than marine geologists can get of their subject matter. Today, however, the more telling comparison is with Venus, a planet whose surface was until recently hidden by thick, unbroken clouds of sulfuric acid. In 1992 a U.S. spacecraft named *Magellan* completed its mission of mapping Venus by cloud-penetrating radar. It surveyed 90 percent of the Venusian surface at a resolution that allows planetary scientists to discern features smaller than a thousand feet across. Oceanographers are not nearly as lucky. Although they have been at their business this past century and more, they have surveyed less than 10 percent of the ocean floor at the resolution *Magellan* offers for Venus.

But in the past two decades their knowledge has improved dramatically, thanks to two new tools that replace Marie Tharp's inspired guesswork with real data. One of the tools is a satellite radar similar to the instrument carried by *Magellan* but operating on a completely different principle, because radar does not penetrate water. The other tool does operate on the same principle as *Magellan*'s radar, but it is a sonar, and what a sonar—the apotheosis, in a way, of Charles Bonneycastle's dream. Both tools were originally developed for the U.S. Navy.

When the Navy launched Geosat late in the Cold War— in 1985, to be precise—its purpose was not to map the seafloor. It purpose was to survey the altitude of the sea surface, all over the world, to within a few inches. The point was not to measure waves, either. Even a calm ocean is not flat: it has hills and valleys that depart by as much as a few hundred feet from what we think of as sea level. The slopes of these features are so gentle—they extend over tens or even hundreds of miles—that no ship ever feels them. Yet the Navy decided that submarine commanders, of all people, would benefit from precise measurements of this imperceptible oceanic topography.

Why? Because the study of bumps on the ocean surface is a reliable kind of phrenology: it reveals deeper truths about the ocean. Small, shifting bumps are created by the shifting fronts between water masses—between the warm Gulf Stream and the cold Atlantic, say—and those same fronts scatter sound, thus creating sonar shadows that can hide an enemy submarine. The larger and more permanent hills and valleys are created by something else entirely: by Earth's gravity field, which is slightly stronger in some places and weaker in others, and does not necessarily point toward the center of the planet. The sea surface bends with these subtle variations, always remaining perpendicular to the direction of gravity, like the bubble in a carpenter's level. Knowing the gravity variations helps a submarine stay on course when it is underwater and sailing blind, and when the time comes to launch a missile at Minsk, that knowledge is essential. If the missile starts out on a slightly wrong heading, it will miss its target, thousands of miles away.

From a near-polar orbit, 500 miles high, Geosat circled the spinning Earth, bouncing a radar beam off the sea surface, painting it with a tight mesh of densely packed radar tracks. The time it took for the radar pulse to bounce back to the satellite revealed its distance from the sea surface, and from those data the Navy could calculate the slope of that surface and the shape of the gravity field. Geosat yielded the most comprehensive set of gravity measurements ever.

For the Navy, the payoff was a better chance of hitting Minsk; and of course that same information, had it been released to the public, could have helped a Soviet submarine hit, say, Cleveland. But not long after Geosat completed its work, the Cold War ended. With it died some of the rationale for keeping Earth's gravity field a secret. In 1995 the Navy was finally persuaded to declassify the Geosat data set. Within months, two civilian geophysicists, David Sandwell of the Scripps Institution and Walter Smith of the National Oceanographic and Atmospheric Administration, had converted it into a map.

That first map was only of the gravity field, but to anyone who had ever seen a map of seafloor topography, its broad outlines looked familiar. That should not be surprising: insofar as mountains tend to have more mass than valleys, topography generates gravity. "If you put a mountain on the seafloor," explains Smith, "the extra material represented by the rocks in that mountain add their own gravity to the overall field. If you're right above the mountain, the added gravity pulls down in the same direction, and so it adds to the magnitude of gravity. But if you're off to one side of the mountain, the gravitational field of the mountain pulls toward the mountain, and so the effect is to change the direction of gravity a little bit." The sea surface, acting as a carpenter's level, follows these changes; it becomes like an attenuated visual echo of the seafloor, piling over mountains, dipping down over trenches.

William Haxby of Lamont-Doherty had demonstrated the principle in the early 1980s with data from Seasat, a NASA satellite—but crudely, because Seasat's ground tracks were never less than 50 miles apart. That is much closer than the lines of sonar data Marie Tharp had to work with, but a large mountain can still slip easily through a 50-mile gap. Geosat's tracks were between 2 and 4 miles apart, and it could resolve features down to about 7 or 8 miles across, especially when Sandwell and Smith supplemented its measurements with data from a similar European satellite called ERS-1. Moreover, by 1997 they had found a way around a basic limitation of a gravity map: the fact that gravity does not *always* follow seafloor topography. The density of seafloor rock affects it too.

Sandwell and Smith's solution was to calibrate the gravity data wherever they could with real depth measurements made by shipboard sonars. By seeing what the relation was between gravity and seafloor topography where the topography was known, they could use gravity to predict uncharted topography.

The map they published in 1997 shows a lot that was not on the charts before Geosat flew. The Foundation Seamounts, for instance: a 1,000-mile-long chain of undersea volcanoes, some soaring 10,000 feet above the surrounding floor of the South Pacific. About half the seamounts on Sandwell and Smith's map, in fact—thousands of mountains that would be eminently skiable if they happened to be on land—were unknown before. Only now, at the close of the twentieth century and long after most people thought the age of exploration was over, is it possible to say that any peak on Earth taller than around 5,000 feet has been discovered.

The presence of all those seamounts means the ocean on average is not as deep as had been thought. The seafloor does indeed cool and subside as it spreads away from the midocean ridges, but apparently it is not a monotonous process. As it moves, it passes over hot spots: slender plumes of hot rock rising from deep in the mantle, hundreds of miles below the surface and completely independent of the overlying plates. All the world's oceans are speckled with isolated volcanoes pushed up by such plumes as the plates pass over them; the Hawaiian Islands are the most famous example. Sandwell and Smith believe the random reheating of the seafloor by hot spots explains why it does not cool and sink as rapidly as had been thought.

A more striking feature of the map are the fracture zones that streak across the ocean basins. Those frozen tracks of the drifting continents were never terribly prominent on the Heezen-Tharp drawings, but on the Sandwell-Smith map they stand out. Look under the bulge of West Africa, and you can see at once how that continent has motored away from South America over the past 110 million years—"motored," that is, at a speed of an inch or two a year. Look in the southern Indian Ocean, and you see the history of Australia's drive for independence from Antarctica; its halo of fracture zones seems almost to be hurling it into the confusion of the western

Topography of the Ocean Floor

Based on satellite gravity altimetry and shipboard depth soundings

Image courtesy of Walter H. F. Smith, National Geophysical Data Center,
National Oceanic and Atmospheric Administration

(For an online color image of this map, as well as other information about it, see
http://www.ngdc.noaa.gov/mgg/fliers/97mgg03.html.)

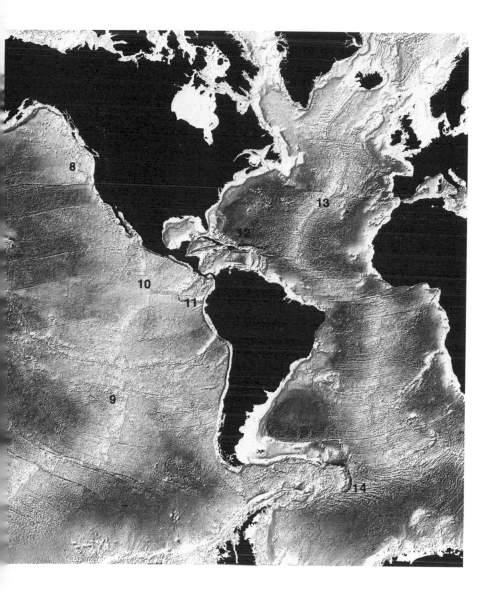

1. Southwest Indian Ridge	8. Juan de Fuca Ridge
2. Carlsberg Ridge	9. Foundation Seamounts
3. Southeast Indian Ridge	10. East Pacific Rise
4. Philippine Trench	11. Galápagos Rift
5. Japan Trench	12. Puerto Rico Trench
6. Mariana Trench	13. Mid-Atlantic Ridge
7. Aleutian Trench	14. South Sandwich Trench

Pacific. Farther to the northwest, the striated floor of the Arabian Sea shows the trail left by India as it plowed into Asia and raised the Himalayas. The fracture zones show the map for what it is—a still from an action-filled video that has been running for hundreds of millions of years, with no prospect of an end.

That is the sort of big picture you cannot get from survey ships crawling over a vast ocean at a few knots. And in the vast tracts of ocean that have never been surveyed, the satellite map offers a first crude glimpse of reality. "This map is going to focus our attention on some places where we had not usually gone with ships, because they're in remote areas in the Southern Ocean, they're far from ports, and the weather down there is uncomfortable," says Smith. "Those areas probably hold the key to how the plate-spreading system actually works. It's having this global view that is really so exciting. We're going to have to rethink all our hypotheses that were based on limited knowledge of a few easy-to-reach places in the Atlantic and the Pacific."

And yet, good as it is, Sandwell and Smith's map still does not provide as good a view of Earth as *Magellan* did of Venus. *Magellan* had a great advantage: Venus is dry. Thus the spacecraft could bounce its radar beams directly off the Venusian surface and make pictures that revealed features as small as a few hundred feet across. On the Geosat map, a seamount smaller than seven or eight miles across would be a blur at best, while two mountains closer together than that would fuse into one. In principle another gravity-mapping satellite might one day do better. But the physical limit is a resolution of about two or three miles, which not coincidentally is also the average depth of the ocean. If two seafloor hills are only two to three miles apart, the bulges their gravity produces at the sea surface will inevitably merge and make them indistinguishable. "You can't beat that limitation," says Sandwell, "unless you drain the oceans."

Or unless you use sound, which, unlike radar, travels through water. The Navy never intended to map the ocean floor with Geosat. Not that the Navy did not care about having such a map; on the contrary, if it was vital for a submarine commander to know precisely the strength and direction of gravity before launching a long-range missile, it was even more vital for him to know exactly where his submarine was. With an accurate seafloor map he could figure that out, even when the vessel was submerged, but a map that showed nothing smaller than eight miles across was not nearly accurate enough. To map the parts of the ocean it cared about, the Navy turned instead to a new type of sonar. That device then went on to transform the study of marine geology.

———

Multibeam sonar, as it is called, was invented in the early 1960s, but for nearly two decades the Navy kept it under wraps. It was invented by a physicist named Harold Farr and by an uncredentialed sonar wizard named Paul Froelich. Farr and Froelich worked in Massachusetts for the Harris Anti-Submarine Warfare Division of the General Instruments Corporation. One day in the late 1950s, Wilbur Harris, their boss, was talking with a couple of men from the radar division about possible joint efforts. One of these men, Arthur Rossoff, brought up a technique that radio astronomers were using to get a clearer look at distant galaxies—a technique called cross-fanned beams. Rossoff suggested that Harris might adapt the technique to sonar for the purpose of getting a clearer look at the ocean floor. Harris suggested that Farr might look into that idea. Farr was not much interested in the ocean floor per se. "I'm not much of a sailor," he says. "I get my kicks out of designing." He and Froelich proceeded to make Rossoff's idea real.

At the time, conventional echo sounders had two main limitations. One was that they produced only a single line of depth measurements beneath a ship. The other was that those measurements were not terribly sharp: the sonar ping spread out over an area of seafloor several miles across, and so the echo and the resulting depth measurement came from some point in that area—which point one could not be sure. Farr's approach was to solve the first problem by turning the second to his advantage.

First he replaced the single sound transmitter of the echo sounder with a line of many transmitters arrayed along the keel of the ship from bow to stern. The sound waves fanned out in a wide swath to port and starboard, but to the fore and aft, they interfered with one another destructively; a sound-wave trough from one transmitter canceled a peak from the next. The result was a fan of beams that "insonified" a long, narrow stripe of seafloor, one that extended for miles to either side of the ship but only a few ship lengths in the direction the ship was traveling.

Next Farr ran a line of receivers *across* the keel, perpendicular to the line of transmitters. An echo that came from directly beneath the ship would reach all the receivers at about the same time, and so it would produce a nice, clear signal when their electronic outputs were added together. But the brilliance of Farr's invention lay in how it handled the echoes that came from off to the sides. An echo arriving from some point off to port, say, would reach the port end of the receiving line first. By measuring that travel time, Farr's system could—just like a conventional echo sounder—determine the distance to the point on the seafloor that had sent back the echo. The travel time alone, though, could not reveal how much of that distance was depth and how much was range to port. But by also measuring the small delay before the echo reached the second receiver in the line and then the third, Farr's invention could determine the angle to the echo source—thus pinning down, by simple geometry, exactly where it was.

Farr's system measured the time delays by electronically

compensating for them, adjusting the signals from the receivers so they would all be in phase and add up to a single, booming echo. It did this after each ping of the sonar, for each echo coming from a different point on the seafloor, one after the other, as if it were tuning into different stations on a radio. Selecting distinct beams of sound from the jumble of echoes, it converted each sonar ping into a swath of soundings. That is why the invention was called multibeam sonar or swath-mapping sonar. The first system Farr built mapped a swath about as wide as the ocean underneath the surveying ship was deep—two or three miles, typically—and measured the depth of the seafloor at 61 points within that swath.

The system would not have worked without a computer to process the data. As it happened, in the early 1960s, not long after the invention of the microchip, computers were just becoming small enough and cheap enough to take to sea. The other invention that dovetailed nicely with multibeam sonar was satellite navigation. Until the 1960s, it was possible for oceanographers to discover a mountain on the seafloor and then lose it again. Navigation in the open ocean was not a precise science; a ship might think it had recorded a sonar echo at latitude x and longitude y, when in reality it had drifted dozens of miles downwind or downcurrent from its calculated position. Today that no longer happens. Modern research vessels navigate not by fickle sun and stars and dead reckoning but by satellite. The modern system of satellite navigation, called the Global Positioning System, consists of some 21 satellites on fixed and regularly spaced orbits, such that from any point on Earth one can receive radio signals from at least three or four of them. By homing in on these signals, a ship can calculate its position to within a few hundred feet or less. In the early 1960s, the Navy was just putting in place the forerunner of the system.

Although the Navy had not asked for multibeam sonar, it responded enthusiastically when the idea was dropped in its lap by General Instruments. For a ship equipped with multi-beam sonar and satellite navigation, mapping the seafloor became like mowing the lawn: one could map back and forth across a region in wide swaths, knowing (thanks to the satellites) that one was neither leaving large unmowed gaps nor mowing the same seafloor twice. By 1965, three Navy survey ships were systematically mapping regions of military interest; in later years there would be as many as 12 of them. By the early 1990s, according to an oceanographer who has seen the data, the Navy had mapped around a quarter of the world ocean. But not many people have seen the data, which are locked away at the U.S. Naval Oceanographic Office in Bay St. Louis, Mississippi. The Navy has not yet seen fit to release them.

Fortunately oceanographers have been able to do more than just drool. The cat began to crawl out of the bag in 1973. When French and American oceanographers organized a joint expedition to the region of the Mid-Atlantic Ridge at latitude 36 and a half degrees north, the Navy allowed them to use its maps of that region. Finally, in 1981, the Navy relented and permitted General Instruments to install a less accurate multibeam sonar—with 16 beams instead of 61—aboard a U.S. research vessel, the Scripps Institution's *Thomas Washington*. Since then similar but improved versions of this system, called Sea Beam, have been installed on a dozen other oceanographic research ships.

By the time marine geologists got Sea Beam, they had already invented for themselves a second type of swath-mapping sonar, one with a specifically geologic purpose. It is called sidescan sonar. The pioneering work was done at the British Institute of Oceanographic Sciences at Wormley in Surrey, which built a device named GLORIA (a silly acronym not worth explaining). Whereas Sea Beam measures only the travel time of sound bouncing off the seafloor, GLORIA measures the varying strength of the echo, the way a camera records on film the varying intensity of reflected light. The result is not a topographic map but a sound image of the seafloor— the oceanographic equivalent of an aerial photograph. Side-scan images capture geologically important information that Sea Beam misses, such as faults that cause only minor changes in seafloor depth. More generally, they show the texture of the seafloor—whether it is rough or smooth, whether it is made of jagged rock or rippled mud.

With the development of swath-mapping sonar, it has become possible for the first time to conceive of completing the explo ration of our planet, by mapping the entire ocean floor, if not at a scale suitable for hiking maps, then at least at one com-parable to that of *Magellan*'s map of Venus. In fact, the U.S. Geological Survey and NOAA have already mapped much of a region that is of greatest commercial interest to the United States: the Exclusive Economic Zone, which extends 200 nau-tical miles offshore. The EEZ covers 3.4 million square nau-tical miles (4.5 million square statute miles, or 2.9 billion acres), about 30 percent more than the land area of the United States and its territories. Government researchers have com-pleted GLORIA maps off all the coastal states and also off some of the territories—huge circles of seafloor around lonely Pacific islands that the United States claims for its own. They have gone over part of the same ground with Sea Beam, making topographic maps to complement the geologic ones.

The mapping of the Exclusive Economic Zone has dem-onstrated just how crude our knowledge is, even of the part of the ocean floor that is right next door. In 1984, a GLORIA survey off southern California revealed a hundred uncharted seamounts, or undersea volcanoes, in a region that had been thought to be flat. Off Hawaii, on the other hand, hills that had been taken for seamounts, hills as much as 2,500 feet high, turned out to be something else entirely: huge blocks of debris from giant landslides that had roared down the sub-merged flanks of the Hawaiian volcanoes and come to rest as much as 60 miles offshore. Similar blocks lay at the mouth of Zhemchug Canyon, off the coast of Alaska in the Bering Sea; apparently Zhemchug, one of the largest submarine canyons in the world, was formed when those blocks broke free from their original position and tumbled down the continental slope. In short, the features that are revealed by swath mapping the seafloor are not trivial details. They are the size of mountains.

Yet there are no plans to map the entire ocean floor, although as big science goes the cost would not be exorbitant. To map the entire ocean with the latest Sea Beam system would cost in the neighborhood of $1 billion. A survey with GLORIA would be cheaper, because GLORIA cuts much wider swaths than Sea Beam. But in any case $1 billion is not an extraordinary sum to spend on mapping our planet. We spent more than half that to map Venus, and we are committed to spending many, many billions on the International Space Station. But the ocean is not space, and the chances of our government funding a large seafloor mapping program are vir-tually nil.

Even if the money were miraculously to appear, though, there would remain a major obstacle to mapping the whole ocean floor. The obstacle is the size of the ocean; it is still big, even when taken in wide swaths. *Magellan*, screaming over Venus at as much as five miles per second, could map nearly the entire planet in less than a year. But oceanographic ships on Earth travel at less than a thousandth the speed—no more than about 15 knots. To map the topography of the entire ocean floor with the most advanced Sea Beam system would take about 200 ship-years—200 ships working nonstop for one year, or one ship working for 200 years, or, more realistically, some intermediate combination. There are 13 civilian ships in the world today equipped with some form of multibeam sonar; if they all devoted themselves exclusively to a systematic mapping of the ocean floor, it could probably be done in about 20 years. In other words, it would be a long project.

And to some marine geologists, anyway, it would be a silly one. They are not interested in mapping the seafloor just to have done with it; they are interested in understanding how it was made. Bill Ryan, Heezen's former student, is a good example. He has devoted a large part of his career to developing tools for seafloor mapping; he helped design a sidescan sonar similar to GLORIA, but capable of taking much finer pictures of smaller areas of the seafloor. Ryan has no desire to map the whole thing once and for all. The notion seems to strike him as a little vulgar. "There's technically no point," he says. "That's not a good way to spend your money. The fact is we can learn how this planet works by seeing five percent of its surface. The next 95 percent looks like the first five percent, it's just rearranged. The human intellect is such that you don't have to see it all to understand it."

"It's sort of like music," Ryan goes on. "You listen to Beethoven or Mozart, and in thirty seconds you know which it is, you have an essence, and you have a warmth, and you're transformed. You haven't listened to the whole symphony, but the content of the symphony, its mood, its texture, its vibrancy, its emotion are there. And it's wonderful sitting for the next hour or half hour for the whole concert, but you can tell Haydn from Mozart from Beethoven with a glimpse of it. And that's basically what we're trying to do: to see if we can tell seafloor created in different tectonic environments, different settings. I would argue strongly that we don't have to map the whole. Why do it?"

Ryan's way to grasp the essential music of the seafloor is to map "type regions": different classes of seafloor landscape, the marine counterparts of alps, tundra, desert, and so on. There are various types of rift valley, of continental canyon, of seamount, of abyssal hill. Whereas the Heezen-Tharp map had only a dozen or so different types of landscape, Ryan thinks you would need a couple of hundred to characterize the sea-floor adequately. He would map as little as one example of each type with multibeam sonar—because when you have seen one, you have seen them all. And whereas Heezen and Tharp had to fall back on their intuition to fill in the huge blanks of uncharted terrain and create a portrait of the whole ocean, Ryan would rely on a computer. The computer would fill in the spaces between the sonar surveys with landscapes of the appropriate type—the right roughness, amplitude, and so on. The map would be realistic, but like Tharp's map, most of it would be unreal. "It would be a lot better than Marie did," says Ryan.

Indeed, even trained marine geologists cannot always distinguish one of Ryan's computer-generated maps from the real multibeam thing. Still, mimicking something perfectly is not the same as understanding it. Understanding the seafloor means knowing how it was created in the first place. And so today, although marine geologists have no hope of seeing a detailed map of the entire ocean floor in their lifetime—that joy (or chore, depending on your point of view) will be bequeathed to our children or more likely to their children—they are nevertheless hoping to map the most important part of it. The seafloor is like a human being, in that its character is determined to a considerable extent at birth. The birthplace of the seafloor, and the place where marine geologists these days are focusing most of their attention, is the midocean ridge.

———

Fill a large flat pan—the size of a cookie sheet but a little deeper—with candle wax. Put the pan on a hot plate, so the wax melts. Now set up a fan that blows across the surface of the wax, cooling it and causing the surface to congeal into a sort of frozen floating scum. Finally, stick a plywood paddle vertically through the wax, and attach it to a small electric motor that will move the paddle toward one side of the pan. Watch what happens.

A seismologist named Jim Brune stumbled on the idea for this experiment one Christmas, when he was making candles with his kids. A marine geologist named Ken Macdonald has pursued the idea. Macdonald is a leading expert on midocean ridges. "Ken has the best insight of anybody about how ocean ridges work," says Bill Ryan. Brune's wax experiment has been the source of some of Macdonald's insight: It is a kitchen-physics model of a midocean ridge.

The layer of frozen scum on the wax is equivalent to the ocean crust and part of the underlying mantle—to a layer of rock between 6 and 60 miles thick called the lithosphere, which has cooled enough to become rigid. The lithosphere is cooled by the water of the ocean—or in the wax model, by the fan. Below the lithosphere lies the asthenosphere, a part of the mantle that is hot and therefore plastic, like the wax at the bottom of the heated pan. Hot rock, like hot wax or hot anything, wants to rise to the surface; that is how volcanoes are born.

A midocean ridge is a long chain of volcanoes, but it is more than that. It is a boundary along which two of the plates that make up the lithosphere, the surface of the solid Earth, are spreading apart. ("Crustal plates," as they are often called, is a convenient shorthand; the plates include a bit of mantle as well as crust.) The plates are pulled apart by gravity. Gravity causes them to slide down opposite flanks of the midocean ridge; and gravity causes the old, cold edge of the plate to sink down into an ocean trench and then into the mantle, dragging the rest of the plate along behind it. In the wax model, both these effects are simulated by the paddle. As it moves slowly toward one side of the pan, it drags the layer of frozen wax along behind it. The wax stretches and cracks, in a pattern that looks a lot like a midocean ridge.

The paddle and the hot plate represent the two fundamental

forces that shape the ocean floor—tectonism and volcanism—and that, coincidentally, divide marine geologists into two camps. Both forces act on land as well, but on land their effects are obscured by a third force: erosion, the steady action of wind and rain, which is always dismantling the structures erected by the horizontal motions of plate tectonics and the vertical motions of volcanism. In the ocean there is no destructive wind and rain, only the gentle blanketing snowfall of sediment. Near a midocean ridge, sediment has barely had time to accumulate.

Near a midocean ridge, then, geologists ought to be able to see the effects of tectonism and volcanism in their purest form, and they ought to be able to distinguish them clearly. That would be true, anyway, if they could take a field trip along a midocean ridge, the way their colleagues on land hike through mountain ranges. But marine geologists cannot do that. They can only drop down onto some isolated peak in a submarine, putter around for a few hours and pick up a few rocks, and never see more than 50 feet in front of their faces. As a result they have had a hard time agreeing on which of the two forces, tectonism or volcanism, is primarily responsible for shaping a midocean ridge, and the seafloor landscape in general. Is the speed at which plates spread apart the main thing, and thus the degree to which they are stretched and pulled into trenches? Or is it the heat of the underlying mantle, and thus the supply of magma to the ridge? In one sense it is a chicken-and-egg question, because everyone agrees that both tectonic stretching and volcanic eruptions happen at midocean ridges. Yet in this case the question of which comes first is not a meaningless one.

On the basis of multibeam sonar maps, wax models, and all sorts of geophysical data, Macdonald puts tectonics first. Stretching the seafloor, he says, is like popping the cap off a soda bottle: it releases the pressure on the rock in the underlying asthenosphere, 20 miles down or more. Because that rock is hot, it begins to rise. As it nears the surface, the pressure on it drops still further, and since pressure is what has kept it solid in spite of its great heat, the rock begins to melt. The molten fraction separates out and rises to the top like cream in milk, forming a lens of magma that floats atop a much larger bulb of partially molten rock. Oceanographers have actually "seen" such a magma chamber a mile or so underneath the East Pacific Rise. By using air guns to make sounds loud enough to penetrate the seafloor, they have recorded the strong echo produced by the molten rock.

The magma chamber under a midocean ridge is no more than a mile or two wide, but it is the smithy in which the entire seafloor is forged, along with the underlying ocean crust. As the plates separate, partially molten rock in the lower part of the chamber slowly cools and solidifies, forming a coarse-grained rock called gabbro. From the thin molten lens at the top of the chamber, sheets of magma shoot upward through fissures in the crust. Some freeze in place, forming slablike vertical intrusions called dikes; some rise all the way to the surface and flow out onto the surface as lava. The lava freezes rapidly in the 36-degree-Fahrenheit seawater, forming fine-grained basalt. It freezes so rapidly that it can assume bizarre bulbous shapes called pillows. Two or three miles of gabbro, topped by a half mile of basaltic dikes, topped by a few hundred yards of pillow basalts—this is the layer cake of rock known as the oceanic crust. Oceanographers have seen these layers with their own eyes, diving in submersibles to places on the midocean ridge where fracture zones have sliced open the crust and exposed it in full cross section. The Vema fracture zone, at 10 degrees north on the Mid-Atlantic Ridge, is one such place; another is the Hess Deep, a three-mile-deep chasm west of the Galápagos near the East Pacific Rise.

What researchers have not seen, with their own eyes or otherwise, is a wall of lava erupting without interruption along thousands of miles of seafloor. On the contrary: In the 15 or so years since multibeam sonar became available, it has revealed that midocean ridges are everywhere broken into discrete segments that interrupt and offset the crest. The longest segments have long been familiar, because they are bounded by the largest offsets—the transform faults. But once their ships had Sea Beam, Macdonald and other researchers began discovering that each of these long segments was itself punctuated by a much finer segmentation—by places where two adjoining stretches of ridge overlap without quite meeting, and by places where the straight line of the crest is not actually interrupted, but only curves a bit. What is more, the shape of the ridge—it undulates up and down like the back of a sea serpent—has turned out to be directly related to these horizontal excursions. The ridge crest is fattest and highest (meaning the water is shallowest) in the middle of a segment and skinniest and lowest at the edges, where the crest curves or is fractured.

The fat places, it seems clear, are places where the ridge is swollen by a bulging magma chamber. Why does magma well up at those particular places? To Macdonald, it is mostly because the seafloor just happened to stretch that way as it was pulled apart by the sinking of the plates, thousands of miles away. The magma rises toward those points that have been stretched most, where the pressure is lowest. With wax models, Macdonald has been able to reproduce the segmented pattern of the ridge. The hot plate alone is not enough to crack the wax; it has to be pulled apart by the plywood paddle. And then it does not crack in a continuous line. It cracks in segments, like the seafloor.

In both wax and rock, the segments are not permanent features; they evolve. They even duel. As the pulling continues, neighboring segments grow toward each other, miss each other, and curve toward each other like wary rivals. The one that is longer to begin with usually outflanks its neighbor, links up with it in mid-segment, and cuts off the outflanked tip.

That creates a continuous stretch of ridge, until the next bout of stretching and cracking segments the ridge once again.

Each time the seafloor cracks, according to Macdonald, each time the motion of the plates stretches the frozen lava in some segment of the rift valley to its breaking point, the magma chamber under that segment may cough out a fresh supply of molten rock, and thus a new chunk of seafloor. The magma balloon itself helps extend the crack to either side, ripping open the seafloor the way Superman's swelling chest rips open his shirt. Each segment of midocean ridge, then, is produced by a feed back between seafloor cracking and magma welling up. But it is the cracking, the tectonic force, that comes first, in Macdonald's view—cracking induced by the motion of the plates.

Not everyone agrees with him. Some marine geologists believe that the volcanic upwelling comes first, that it and not the tectonic motion of the plates is what causes the seafloor to crack. The volcanists have a bit of almost-kitchen physics on their side to counter Macdonald's wax models. In the early 1980s, a fluid dynamicist at Woods Hole named Jack Whitehead did a series of experiments designed to simulate what happens when one fluid is trapped under another that is more dense and more viscous—specifically, when a hot and ductile asthenosphere is trapped under a cold, dense lithosphere. Whitehead injected a straight line of water and glycerine—a syrupy liquid, used in antifreeze, that is denser than water—into the center of a bath of pure glycerine. Not surprisingly, the water-glycerine mixture, being less dense than glycerine alone, started to rise to the top. What was more striking was how it rose: in individual balloonlike bulbs spaced at regular intervals along the water-glycerine line and tethered to the line by long, slender stalks. If one thinks of the water-glycerine mixture as the asthenosphere, then each of those bulbs becomes a magma chamber under its own segment of midocean ridge. And the asthenosphere, or at least a region of it that for some reason is especially hot, becomes buoyant enough to crack the overlying plate, sort of like a chick poking through an eggshell.

Does the real asthenosphere act that way? Are midocean ridges places where the lithosphere is being pulled apart or places where it is being melted from below? Which comes first, the chicken or the egg? Oceanographers still do not know for sure.

Nor do they know for sure why different midocean ridges are so radically different in appearance. The Mid-Atlantic Ridge and the East Pacific Rise represent the extremes of a spectrum. The Atlantic one is the archetype, the one that Heezen and Tharp knew: a ridge with relatively steep flanks, cut along its crest by a rift valley 1 to 2 miles deep and 10 to 20 miles wide. The East Pacific Rise is another beast entirely, which is one reason that Heezen initially had trouble convincing West Coast oceanographers that his globe-girdling rift was real. The East Pacific Rise is a gentle swelling of the seafloor, typically a thousand feet high or so; it could fit comfortably inside the Atlantic rift valley. And the rift on the crest of the East Pacific Rise is not a proper valley at all: no more than a few hundred feet deep and a mile or so wide, it was all but undetectable with the sonar systems of Heezen's day.

The difference between the Atlantic and Pacific ridges must somehow reflect their different histories. The Pacific is the older of the two oceans. It is a remnant of Panthalassa, the global ocean that surrounded the supercontinent Pangaea until about 170 million years ago. That is when the supercontinent began to separate into the continents of today, along a rift that became the Atlantic. As the plates bearing the continents have scattered away from the Atlantic rift, they have overridden the floor of the Pacific; the entire Pacific is rimmed with deep trenches, where the ocean floor is plunging under the continent-bearing plates and back into the mantle. The trenches are destroying seafloor much faster than the East Pacific Rise is creating it, and so the Pacific is shrinking.

Yet though it may at first sound paradoxical, the East Pacific Rise is creating new seafloor much faster than the Mid-Atlantic Ridge is. The East Pacific seafloor is spreading apart as much as nine inches a year, whereas in the Atlantic the spreading rate is closer to one inch a year. According to Macdonald, the difference lies in the Pacific trenches: the plates are spreading apart so fast along the East Pacific Rise because they are being pulled that fast by the trenches. The Atlantic seafloor is being destroyed only in two small places—the Puerto Rico Trench and the South Sandwich Trench off the tip of South America—and so it spreads slowly.

Besides their spreading rates, the Pacific and Atlantic ridges are different in another way: The East Pacific Rise has a far richer and more reliable supply of magma. Although volcanic eruptions certainly happen along the Mid-Atlantic Ridge, seismologists have yet to detect any evidence for a magma chamber underneath it, like the one that exists all along the East Pacific Rise. The magma supply in the Atlantic seems to be a sputtering one at best. The exceptions are the hot spots: the Atlantic ridge happens to lie on or near several of those rising plumes of hot mantle, which add their strength to the ridge's own shallower and fickler source of magma and raise its peaks above sea level. Iceland is a massive volcanic construct of this type; the Azores and Ascension Island are smaller ones. The Mid-Atlantic Ridge is at its most active just south of Iceland. When Bruce Heezen died in the NR-1 submarine, he was headed to that portion of the ridge, in the hope of seeing an eruption.

Except at the hot spots, however, volcanic activity in the Atlantic is more sporadic than in the Pacific. Macdonald and others of the tectonic school would say that is simply an effect of the difference in spreading rates: magma wells up more continuously along the East Pacific Rise because the seafloor is cracking more continuously and taking the lid off the asthenosphere. Ryan and other volcanists would argue that there is more magma under the East Pacific Rise because the mantle is hotter there. Either way, the difference in magma supply helps explain why the East Pacific Rise and the Mid-Atlantic Ridge are such different places.

So much magma wells up in the East Pacific that it is always filling the rift, which is just a shallow volcanic caldera, a trough formed where a pool of lava has drained away. Anyway, trying to dig a deep valley under the East Pacific Rise would be like trying to dig one in soupy mud—the crust is so hot and plastic that the sides of the valley would simply flow in. Under the Mid-Atlantic Ridge, on the other hand, there is not enough magma to fill the gap left by the spreading plates, and the crust is cold and rigid almost all the way through its six-mile thickness. Thus as it is pulled apart it can break to great depth along sloping faults that flank the axis of the ridge. The block of crust inside these "normal" faults slides down the fault planes, forming a deep valley. The blocks outside the faults rise up along them, buoyed by the fluid mantle that has been displaced by the falling crust. Those outside blocks become the tall mountains that flank the valley.

On the flanks of the mountains, or of the gentle rise in the case of the East Pacific, is where the abyssal hills begin. They do not end there; the spreading plates carry them out over the ocean floor, and the hills disappear only when they are covered by sediment or subducted in a trench. Abyssal hills are the single most common geologic feature on Earth; they cover something like a third or half of the planet. When that fact was discovered in the early 1950s, it came as a great shock: the seafloor was not, after all, essentially flat. Volcanists and tectonists have now been arguing for more than 40 years about how the hills are made. Are they just piles of lava that erupted at the midocean ridge and were then rafted away on the spreading plates? Or are they smaller versions of the Mid-Atlantic Ridge itself—a series of high-standing "horsts" formed when the stretching of the crust allowed the blocks between the hills to subside along normal faults?

The answer depends again on which midocean ridge you visit. In some places whole volcano cones do seem to form near a ridge axis and then drift away on one of the plates. At the Juan de Fuca Ridge, a small spreading center off the coast of Washington State, Ryan and his colleague Ellen Kappel discovered something even stranger: split volcanoes. In a number of cases, they found that a pile of lava had accumulated directly on the ridge axis and had then been split in two as the plates spread apart. On opposite sides of the ridge Ryan and Kappel found matching volcano halves that formed mirror-image abyssal hills. "The seafloor is a predominantly volcanic landscape," Ryan says. "Abyssal hills are created at the ridge, and they're frozen in as the seafloor comes out."

But not at the East Pacific Rise, according to Macdonald. In 1994, he and his colleagues made a series of 17 dives in the submersible *Alvin* onto one part of the rise. Across the axis they puttered, video cameras whirring and cameras flashing, up one flank and down the other, and then along the flanks too. They also dropped onto "mature" abyssal hills, more than 20 miles from the rise axis, that were 700,000 years old. In between those hills and the axis they could see hills that were still being born. And what they saw were horsts, or rather horsts in volcanic clothing. The inner slopes of the hills, the ones facing the axis, were near-vertical cliffs, some as much as 500 feet high, which is just what one would expect from a normal fault. The outward-facing slopes were faults too—but they were exposed in only a few places, where Macdonald and his colleagues found gaps in the gentle folds of frozen lava that had cascaded down from the crest of the hill and draped themselves over the fault scarp.

Hills like that could be mistaken for split volcanoes, but

really they are horsts in disguise. Showing no interest in resolving the great volcanism versus tectonism debate, the seafloor had simply found a way of making things confusing.

A lot of things remain confusing and uncertain about the midocean ridge. But if one thing has become clear, it is that the birth of the seafloor is fitful, episodic. The old diagrams of plate tectonics, the ones that show sheets of red magma welling up along the whole of the ridge, are seriously misleading. The ridge's segmentation is central to its nature, and it is a segmentation in space as well as time. The 200- to 400-mile-long segments bounded by transform faults retain their identity for millions of years. But they are made of smaller segments that evolve more quickly, with the smallest of all corresponding to individual bursts of seafloor cracking and eruptions. Along any given stretch of the slow-cold Mid-Atlantic Ridge, such bursts may come only once every 100,000 years. Along the fast-hot East Pacific Rise, they happen far more frequently—every thousand years, every hundred, maybe even every ten; no one really knows.

A few years ago, though, Macdonald stuck his neck out about one section of the East Pacific Rise. It lies between the latitudes of 9 and 10 degrees north, about 1,300 miles due west of Costa Rica, and it was one of the first places he visited when he started using Sea Beam in the early 1980s. The Sea Beam maps showed that the rise in that area had an unusually broad, swollen shape; it looked decidedly pregnant. A few years later seismologists discovered that it was underlain by a rich magma chamber. In 1988 Macdonald and Jeff Fox of the University of Rhode Island issued a forecast: the East Pacific Rise between 9 and 10 degrees north, they said, was ripe for an eruption. If oceanographers were lucky, they might witness what none of them had seen before, what Heezen had been after when he died—a new piece of seafloor being born.

As it turned out, Macdonald and Fox were right. And the events that transpired at 9 degrees, 50 minutes north would later remind Macdonald of something that had happened to him two decades earlier, on one of the earliest cruises he had been on as a young oceanographer. That was in 1972, just a few years after his decision to forgo medical school and a secure existence for a career in marine geology. Plate tectonics was hot then, and new and exciting; and Macdonald had told himself that there had to be more to say about midocean ridges than just that plates spread apart along them. All the details of seafloor creation remained to be discovered. The project seemed worthy of a career. In 1972, Macdonald was cruising northeast of the Galápagos, above an offshoot of the East Pacific Rise called the Galápagos Rift. He was trying to see whether he could use sonobuoys, devices the Navy uses to listen for submarines, to detect the rumble of earthquakes on the ridge, and thus to determine when it was active. He found that he could.

Then one day, Macdonald recalls, "We were steaming around doing something else, and we noticed that there were a lot of dead fish at the surface. And we said, 'God, this is really strange.' The truth of the matter is, we've never seen anything like it since. So we netted a few of the fish and brought them back to an ichthyologist at Scripps. He said that they lived at great depth—at the depth of the ridge. They had also died recently—some were still alive at the surface. He did an autopsy and he couldn't pin down what killed them. But in retrospect we think it must have been related to earthquakes I was seeing with the Navy sonobuoys."

In retrospect it was clear to Macdonald that something significant had happened on the Galápagos Rift. It was a geological event, first of all, but it was one that would give rise, just five years later, to one of the greatest biological discoveries of the century. From then on that spot on the seafloor, the earthquake epicenter, would be known as the Rose Garden.

CHAPTER 4

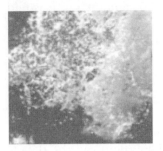

Islands in
the Deep

Surely then everyone must allow that it is quite impossible to comprise every species of terrestrial animal in one general view for the information of mankind. And yet, by Hercules! in the sea and in the Ocean, vast as it is, there exists nothing that is unknown to us, and, a truly marvelous fact, it is with those things that Nature has concealed in the deep that we are best acquainted.

—PLINY THE ELDER, quoted in the *Challenger* Report

The very deep did rot: O Christ!
That ever this should be!
Yea, slimy things did crawl with legs
Upon the slimy sea.

—SAMUEL COLERIDGE, quoted in the *Challenger* Report

O n the floor of the deep sea, the cold and lightless, still and crushing deep sea, every grain of sediment has been through a worm or some other slimy mud eater several times at least. On the floor of the deep sea, where time was once said to pass slowly, if at all, a scrap of flesh falling from above

will be shredded in minutes by frenzied crustaceans and car-
nivorous fish. On the floor of the deep sea, that changeless,
monotonous should-be desert—or so it was once said—the
diversity of species rivals that of the tropical rain forest: we
might find 10 million species of animal or more on the deep-
sea floor were we ever to explore it thoroughly, which, given
its size and our limitations, we have only just begun to do.

These perceptions of life in the deep are relatively new.
They date from the 1960s at the earliest, and mostly they are
more recent than that. Deep-sea biology as a scientific pursuit,
however, is not new. If one ignores ancient sages such as Pliny,
one can say that deep-sea biology began in the early nineteenth
century. (Pliny richly deserves to be ignored on the subject:
his grand total of all species living in the sea, deep or not, was
176—four fewer than Aristotle had found four centuries earlier in the Aegean Sea alone.) In the early nineteenth century,
naturalists first acquired the ability to dredge mud and rock
and conceivably animals from great depth. That was when the
question of what might live on the ocean floor emerged from
the realm of sheer speculation and monster myth and entered
the realm of science. Curiously enough, the science of deep-
sea biology began with the proposition that no animal could
survive in the deep at all. It began with a man named Edward
Forbes.

Forbes was born on the Isle of Man in 1815, and he lived
there, at the bosom of the Irish Sea, until he was 16. He tried
to become an artist but was refused admission to the Royal
Academy in London; went to Edinburgh to study medicine (to
please his mother) but never showed up for the final exam;
and finally returned to his first love, which was marine life. As
a boy on Man, Forbes had collected shells on the beach and
had even done a bit of dredging for bottom dwellers. At Edinburgh, when he should have been cracking the medical books,
he had dredged in the Firth of Forth. By 1839, when he was
24, Forbes was a long-haired philosopher (as some contem-

poraries saw him), with a passion for his subject but no degree and no job. He had become such an accomplished dredger, though, that in that year he managed to convince the British Association for the Advancement of Science to set up a Dredging Committee, consisting primarily of Forbes. The association granted him £60 to dredge creatures from the seafloor. That same year Forbes entertained his friends in the Red Lion Clubbe with his "Song of the Dredge":

> Hurrah for the dredge, with its iron edge
> And its mystical triangle,
> And its hided net with meshes set
> Odd fishes to entangle!

Although Forbes's future as a lyricist was to prove limited, the grant from the British Association was his first big break as a dredger. His second came two years later, when he was offered the chance to serve as naturalist aboard H.M.S. *Beacon*, a Royal Navy ship assigned to survey the Aegean. For a year and a half the *Beacon* crisscrossed the Aegean waters. During that time Forbes was able to drag his small, triangular, leather-mesh dredge along the bottom at a hundred stations, at depths ranging from 1 to 230 fathoms (1,380 feet). He collected hundreds of different species of animal, and he saw that they were distributed in eight different depth zones, each containing its own distinct assemblage of animal life, the way zones of elevation on the side of a mountain are populated by distinct sets of plants.

Forbes also thought he saw, as he later told the British Association, that "the number of species and individuals diminishes as we descend, pointing to a zero in the distribution of animal life as yet unvisited." This zero, Forbes speculated casually— he simply extended a line on his graph of animal number versus depth—probably began at a depth of 300 fathoms. Below that was the final zone in Forbes's scheme, zone nine, a zone that covered most of the ocean floor and thus most of the solid surface of Earth: the "azoic zone," where no animal, to say nothing of plants, could survive.

Forbes's azoic zone was entirely plausible at the time, and it was certainly far from the strangest idea that was then entertained about the deep sea. In the first decade of the century a French naturalist named François Péron had sailed around the world measuring the temperature of the ocean, had found that the deeper he looked the colder it got, and had concluded that the seafloor was covered with a thick layer of ice. Péron ignored the fact that water expands when it freezes and that ice therefore floats. A more popular belief at the time was that water at great depth would be compressed to such a density that nothing could sink through it; when a ship sank, people in the early nineteenth century thought, it never reached bottom but instead floated ghostlike through the abyss on the layer of dense water, its unfortunate crew never achieving their last rest. This ghoulish bit of folklore ignored the fact that water is all but incompressible. But even the more temperate naturalists of the day were guilty of a similar misconception: they imagined the deep sea as being filled with a stagnant and undisturbable pool of cold, dense water. In reality the deep is always being refreshed by cold water sinking from above.

The central implication of all these misconceptions was that nothing could live in the abyss, just as Forbes's observations seemed to indicate. But Forbes had been led astray in two ways. One was the particular study site he happened to use as a springboard for his sweeping postulate of a lifeless abyss. Although the Aegean, being Aristotle's local ocean, had been the birthplace of marine biology, its depths are now known to be exceptionally dull—"faunally depauperate," as a zoologist would put it. Moreover, Forbes was not particularly successful at sampling such life as did exist at the bottom of the Aegean. It was his beloved dredge that was inadequate. Its "mystical triangle" was so small and its mesh so coarse that the dredge inevitably missed animals. Many of those it did catch must have poured out its open mouth when Forbes reeled it in. His azoic zone, then, was a plausible but wild extrapolation from pioneering but feeble data.

As it turned out, the existence of the azoic zone had been disproved even before Forbes suggested it, and the theory continued to be contradicted regularly throughout its long and influential life. Searching for the northwest passage from the Atlantic to the Pacific in 1818, Sir John Ross had lowered his "deep-sea clamm"—a sort of bivalved sediment scoop—into the waters of Baffin Bay, which he determined to be more than a thousand fathoms deep in some places. Modern soundings indicate he overestimated his depths by several hundred fathoms, but in any case Ross's clamm dove several times deeper than Forbes's dredge. It brought back mud laced with worms, and starfish that had entangled themselves in the line at depths well below the supposed boundary of the azoic zone.

In the decades that followed Forbes's "discovery" of the lifeless realm below 300 fathoms, more worms and more starfish were recovered from progressively greater depths. Ships exploring routes for undersea telegraph cables found animals clinging to their sounding lines; and animals sometimes clung to the cables themselves. When a failed cable was retrieved from a thousand fathoms of water off Sardinia in 1860 and was found to be encrusted with animals, doubts about the azoic zone began to spread. Yet although Forbes himself (who died in 1854) had never regarded the zone as anything more than an hypothesis, his followers clung to it tenaciously—like starfish to a sounding line. The theory made too much sense not to be true. What animal could possibly survive in the cold, dark, foodless, and crushing grip of the deep sea? Why believe a few isolated counterexamples collected by amateurs on telegraph survey ships? Those starfish might not have lived on the bottom. They might have "convulsively embraced" the sounding line as it was being hauled in through shallower water.

At the same time that some naturalists accepted the azoic zone unquestioningly, however, the idea was like a thrown gauntlet to others. That was its great virtue: it was an eminently simple, eminently testable hypothesis, and by inspiring expeditions whose goal was to test it, it gave birth to deep-sea biology. The most important such expeditions were those of H.M.S. *Porcupine*, a 382-ton gunboat that the British Admiralty had prevailed on to convert to scientific use and to the cause of exploring the azoic zone. One of the chief prevailers was Charles Wyville Thomson, who later assumed what had been Forbes's chair in natural history at the University of Edinburgh, and who in 1872 was to lead the *Challenger* expedition. In the summer of 1869, Thomson led the *Porcupine* on a cruise off the southwestern tip of Ireland, to the edge of what is now known as the Porcupine Abyssal Plain.

The *Porcupine* had a 12-horsepower engine for hauling in a dredge from great depth. The dredge itself consisted of a mesh bag attached to a rectangular metal frame four and a half feet long by six inches wide. As the dredge was dragged along the seafloor, water, mud, and animals would pour through its metal maw; the water and most of the mud would pour out the half-inch mesh, while the animals settled into the bottom of the bag, which was made of tightly woven canvas. That was the theory, anyway. But during the course of the cruise, Thomson and his colleagues found they needed to make an improvement in the design of their dredge. "In many of our dredgings in the *Porcupine* at all depths," Thomson later recalled,

> we found that while few objects of interest were brought up within the dredge, many Echinoderms, Sponges, and Corals came to the surface sticking to the outside of the dredge-bag. . . .
>
> Finally Captain Calver sent down half a dozen of the 'swabs' used for washing the decks, attached to the dredge. The result was marvelous. The tangled hemp brought up everything rough

and movable which came in its way, and swept the bottom as it might have swept the deck. . . .

On the evening of July 22, 1869, the *Porcupine* mopped the seafloor at a depth of 2,435 fathoms—10 times deeper than Forbes had gone in the Aegean and by far the deepest dredging ever attempted at that time. The dredge took an hour to reach the bottom. The crew left it there for three hours, anxiously watching the rope to see whether it would snap under the tremendous strain. Finally at 9 P.M. they started hauling it in. Later Thomson described the scene to a friend:

> At one o'clock precisely, a cheer from a breathless little band of watchers intimated that the dredge had returned in safety from its wonderful and perilous journey of more than six statute miles. A slight accident had occurred. In going down the rope had taken a loop around the dredge-bag, so that the bag was not full. It contained, however, enough for our purpose—1 ½ cwt. of "Atlantic ooze"; and so the feat was accomplished. Some of us tossed ourselves down on the sofas, without taking off our clothes, to wait till daylight to see what was in the dredge.

One can only imagine the exhiliration at daybreak; Thomson was matter-of-fact. "Life extends to the greatest depths," he concluded, "and is represented by all the marine invertebrate groups." From the 150 pounds of gray, chalky mud, he and his collaborators sifted five species of mollusk, two species of echinoderm, an annelid worm or two, a sponge, numerous single-cell foraminiferans, and more. In the previous three decades Forbes's idea of a bottom limit to deep-sea life had been stretched ever downward to accomodate each new and deeper dredging that brought back living organisms. But now, with a large variety of animals found at a depth of two-and-three-quarter miles, deeper than the average of the whole world ocean, the boundary of the azoic zone could no longer be stretched. Now it had snapped. Now the deep sea was, once and for all, alive; and the idea of an azoic zone anywhere on Earth's surface should have been dead, once and for all.

But biologists continued to kick the corpse, as it were, for nearly a century. The great *Challenger* expedition, which Thomson organized and which was made possible in part by the public excitement over the *Porcupine*'s results, documented that deep-sea life existed all over the world and not just southwest of Ireland. Indeed, one of the ideas Thomson and the others were most interested in testing when they set out on *Challenger* was in a sense the opposite of the azoic zone hypothesis: the idea that most if not all of the seafloor was covered by a pulsing mass of protoplasm called *Bathybius*.

Bathybius had first been "found" by Thomas Huxley, the great evolutionary biologist known to historians as Darwin's Bulldog. At Huxley's request, the surgeon on H.M.S. *Cyclops*, a telegraph survey ship, had collected mud from the floor of the North Atlantic and preserved it in jars of alcohol to bring back to Huxley. When Huxley examined the ooze, he saw it was laced through with a network of slimy, gelatinous material that resembled egg white and that floated free, quivering and shifting around in a lifelike way, when he shook the jar. Embedded in the slime were granules of unidentified material that the slime had apparently eaten. Huxley decided this was no ordinary slime. This was living, animal slime, and "Urschleim" at that: the most primitive possible form of life, pure protoplasm, whose existence had been predicted by Huxley's fellow Darwinian, the German biologist Ernst Haeckel. Huxley named the creature *Bathybius haeckeli*; and Haeckel, delighted, declared that *Bathybius* was the primordial mother of us all, a living ancestor of all living things, still being generated spontaneously out of seafloor mud. It might even be a single globe-spanning organism.

The *Challenger* scientists looked for *Bathybius* all over the globe, however, and all over the world they found plain mud—until one day the chemist on board noticed that some of the mud samples, the ones that had been preserved in alcohol, as Huxley had instructed the *Cyclops* surgeon to do, instead of in seawater, did show traces of the distinctive slime. On analyzing the slime, the chemist found it was neither animal nor vegetable (nor both, as Haeckel had believed) but boring, lifeless mineral; specifically, it was calcium sulfate, which precipitated out of the mud when it was mixed with alcohol. Huxley immediately recanted. *Bathybius* became an historical footnote, albeit a fascinating one.

Certainly it was a strange interlude in the development of deep-sea biology, the main thrust of which—if one can speak of a thrust in a field that proceeded in fits and starts and long hiatuses—was the gradual slaying, by a thousand cuts, of the azoic zone hypothesis that had started it all. One of the last blows was delivered by the Danish ship *Galathea* in 1951. Although Thomson had announced in 1869 that "life extends to the greatest depths" of the ocean, the chief purpose of *Galathea* on its round-the-world expedition was to verify that life did extend to the truly greatest depths—the bottoms of the Pacific trenches. It did: on August 21, 1951, the *Galathea* dredged animals up from the bottom of the Philippine Trench, a depth of about 33,600 feet. The Danish biologists identified eight species of animals—sea anemones, sea cucumbers, bivalve mollusks, crustaceans, and bristle worms.

But the Philippine Trench is not the deepest trench, and so in principle, at least, the azoic zone hypothesis had one last gasp in it. It received its *coup de grace* on January 23, 1960. On that day two explorers accomplished a feat that was less physically taxing, certainly, than Edmund Hillary and Tenzing Norgay's ascent of Everest seven years earlier. It was less difficult technically and dramatic visually than Neil Armstrong and Buzz Aldrin's moonwalk eight years later. Historically, however, it was a comparable milestone, though it is little remembered today. On that day in 1960 a Swiss named Jacques Piccard and a U.S. Navy lieutenant named Don Walsh descended into the deepest chasm on Earth, the Challenger Deep of the Mariana Trench.

They did so in a "bathyscaphe" designed by Jacques's father Auguste, a balloonist who in 1932 had set the record for the highest ascent in a balloon. The *Trieste*, as the bathyscaphe was called, was a zeppelin of the sea—huge, unwieldy, faintly ridiculous looking, and a technological dead end. For the most part it resembled an ordinary submarine, except that what should have been the hull was a series of enormous gasoline tanks and two hoppers containing 16 tons of iron pellets. The iron carried the bathyscaphe into the abyss; when it was released there, the buoyant gasoline lifted the bathyscaphe back to the surface. Piccard and Walsh themselves sat crammed together in a metal sphere, six-and-a-half feet in diameter, that was suspended from the gasoline tanks like a gondola from a balloon.

When they landed on the bottom, 35,800 feet down, after several hours of sinking, one of the first (and only) things they saw out their single porthole was what looked like a flatfish, similar to a sole. Piccard looked at the fish; the fish, or so it seemed to Piccard, looked back. Then the animal bestirred itself and swam slowly out of the *Trieste's* spotlight, vanishing into the black. Piccard and Walsh, too, had to return to their own world after only 20 minutes on the bottom. But the memory of his close encounter with a deep-sea animal stayed with Piccard. "Here, in an instant, was the answer. . . ." he wrote later. "Could life exist in the greatest depths of the ocean? It could!"

By that time, though, no deep-sea biologist would have doubted it anyway. By that time the modern era of deep-sea biology was ready to begin.

Trying to find out what lives on the ocean floor, and how it lives, is hard. Think about what it would be like to study our own world the way biologists must study the deep sea. Imagine a race of space aliens, the kind you read about in supermarket tabloids, but scientifically inclined. Having discovered Earth, they set out to understand the life on its surface. These unfortunate aliens, however, lack tractor beams, teleportation, and other advanced tools of research. Their wispy bodies cannot withstand the crushing pressure of our atmosphere. It is utterly opaque to their eyes, which are sensitive only to a small range of wavelengths in the infrared. For want of a better idea, they decide to blindly drag a large, sturdily framed net from their spacecraft as it cruises safely above the clouds.

The net touches down one Friday evening in Tyler, Texas, where it first clips a flag pole off the top of the courthouse. Next it bounces through a playground in Bergfeld Park, scattering children and parents without snaring a one; collects a dog that is studying a soup bone on College Street; and swoops into a backyard a few blocks later, picking up an azalea bush, a clothesline bearing some exotic black undergarments, and a patch of lettuce with its associated groundhog. Finally it nearly comes to grief in a dark corner of the Sears parking lot, where the weight of a 1979 Chevy Nova and the teenagers in its back seat causes the net cable to groan audibly. The aliens quickly reel in their catch: one quadrupedal carnivore, one quadrupedal herbivore with food, assorted mysterious shell fragments and a large metal-crusted animal (the Nova) whose source of food remains obscure, but which seems to be serviced by remarkably sophisticated intestinal endosymbionts.

This is the approach to deep-sea biology pioneered by the *Porcupine* and *Challenger* expeditions. It is still in use today.

The point is not that it is a bad method; for most of the past century and a half, until the advent of research submersibles, it has been the only method. Moreover, the *Challenger*'s dredgings and trawlings established some important truths about the deep sea. Among other things, they established that, acre for acre, the deep sea is sparsely inhabited: the number of individual organisms is much smaller than it would be in an area of the same size in shallow water. If one were to dredge a square of deep-sea floor five yards on a side, one would typically find only an ounce or two of life in all. The animals are scarce, on average, and they are small, on average. The reason, as the *Challenger's* naturalists recognized, is that there is not much food in the deep sea. No plants can grow there, in the absence of sunlight. The only source of food in most of the deep are the assorted bits of detritus—plant remains, animal remains, excrement—that fall from the surface.

But the *Challenger* data, while they are still cited today, also created some false impressions of the deep sea. One was that life in the deep was, in a way, rather boring. "We got tired on the *Challenger* of dredging up the same monotonous animals wherever we went," H. N. Moseley, one of the naturalists on board, told his audience at a public lecture several years after the ship had returned home. Not only did the number of individual animals plummet as one went from shallow to deep water, but the number of species did as well, according to some interpretations of the *Challenger* data. The deep sea, it seemed, was inhabited by a homogeneous fauna of cosmopolitan species, everywhere more or less the same. This made perfect sense, because the deep sea itself, unlike shallow waters and unlike the land, seemed everywhere the same: everywhere cold, dark, muddy, and short of food. In such an environment, ecologists expected a few hardy species to win out, driving their competitors into extinction. The theory is called competitive exclusion. For nearly a century the belief that the deep sea was a faunally depauperate desert held sway even among the people who loved to study it.

And yet the belief was false. It was proved false in the mid-1960s by two biologists at the Woods Hole Oceanographic Institution, Howard Sanders and Robert Hessler. Sanders and Hessler decided to make a detailed study of one region of the deep sea, along the transect from Cape Cod to Bermuda. They began by dredging mud from the bottom at a series of stations ranging in depth from 4,600 feet, on the continental slope, to more than 15,000 feet, on the abyssal plain near Bermuda. They got a lot of animals—but more important, they could tell that they were not getting anywhere near all of them. Hessler, an expert on a group of segmented crustaceans called isopods (a group that includes pillbugs on land), noticed that he was getting different species of isopods with his dredge than had researchers who had glided over the bottom with a trawl. He was getting more of the burrowing species; they had gotten more of the swimmers.

Hessler and Sanders decided to build a device that would combine features of both a dredge and a trawl—a device that would bring back a better sample of everything that lived in the seafloor mud (the "infauna") as well as on or just above it (the "epifauna"). The "epibenthic sled," as they called it, looked in outline like a tank without a turret. Instead of treads it had two eight-foot-long steel runners that kept it from sinking too deep into the bottom mud. Between the runners Hessler and Sanders suspended a bag made of nylon mesh, with openings one millimeter wide. The mouth of the bag was a foot high, and as the sled was towed along, the steel rim that framed the mouth bit about an inch into the mud. Since most of the organisms that burrow into the mud live near the top, the sled allowed Hessler and Sanders to capture a fairly good cross section of the swimming, crawling, burrowing, and just plain sitting populations of the deep sea.

They and every other deep-sea biologist were astonished at what they found. Pulling the sled along the continental slope for an hour at a depth of 4,600 feet, Hessler and Sanders brought back 10,804 worms of 91 different species; 4,393 arthropods of 127 species, mostly isopods and other crustaceans; 6,713 mollusks of 84 species; and 2,715 echinoderms of 20 species, mostly starfish and sea cucumbers. In all they caught 25,242 individual organisms of 365 species in an hour of trawling at a single site. All the way down in the abyssal plain, at 15,700 feet, their catch was less abundant—a total of 3,737 animals—but still very species-rich, 196 different species of animal in all. Hessler and Sanders had verified one of the tenets of post-*Challenger* deep-sea biology: the animal population on the ocean floor gets sparser the deeper you go. But they had shattered the other one: the deep sea was not faunally depauperate. In fact, its sparse population was incredibly diverse. Hessler and Sanders had found more species on a small part of the abyssal plain than Sanders had found just a few years earlier right off Woods Hole, in the shallow and teeming waters of Buzzards Bay.

How could so many different types of animals manage to survive in such a hostile and apparently uniform environment? Sanders proposed a theory that stood competitive exclusion on its head. The diversity of the deep sea, he said, arose from its very stability: from the fact that its temperature, salinity, and other characteristics had remained unchanged over millions of years. With no sudden catastrophes to drive old species into extinction, the total number of species gradually grew over geologic time, as new species continued to evolve and fill every conceivable ecological niche. In peaceful, unstressful conditions, in a world that never confronted them with the unexpected, the organisms did not so much adapt to their physical environment as adapt to one another, evolving complex food webs that allowed the manifold species to live and let live (although individual organisms continued to eat and be eaten). The deep sea, in Sanders's view, was a kind of ecological nirvana. Nor was it the only one on the planet: Sanders thought his theory could explain the tremendous diversity of the tropical rain forest as well. The rain forest held a lot more food and thus a lot more individual organisms, but it too was a stable place; it had been unfailingly warm and wet for tens of millions of years.

In a sense, Sanders was just putting a positive slant on the same old abyss that had existed in the minds of biologists since *Challenger*, in the sense that nirvana too is a boring place. "Altogether the deep sea, cold, dark, and still, must be about the slowest place to live in that can be imagined," was the way Moseley, the *Challenger* naturalist, had described it. Later in the same lecture he had said: "Life must be very monotonous in the deep sea. There must be an entire absence of seasons, no day and night, no change in temperature. Possibly there is at some places a periodical variation in the supply of food falling from above, which may give rise to a little annual excitement amongst the inhabitants."

Only in that last remark does Moseley now seem prescient. Today the deep sea seems more exciting than it did either to Moseley ("The unhappy deep-sea animals have not escaped their parasites in their cold and gloomy retreat") or even to Sanders. No one really knows just why it is that life in the deep is so diverse; there have been many theories. But the idea that the diversity arises primarily from the stable monotony of the place is no longer fashionable.

It is probably no coincidence that the same change should have happened in our view of the rain forest. From an airplane (one need not go so high as an alien spaceship) the Amazon too looks like a monotonous place—a sea of green, uniform and unchanging. One has to descend beneath the treetops to see the variety. Then it becomes clear that the Amazon is not one place but many places—many "patches," to use a term that is now popular among ecologists—all of them different. It is a patchwork in space but also in time: The boundaries of the patches are constantly shifting, and within each one, there is a lot going on. A storm sweeps through the forest, a river jumps its banks or erodes them till they crumble, a great sun-blocking tree falls and dies. None of these disturbances is enough to eradicate a species, but each one kills some individual organisms and in so doing makes room for others, possibly of a different species. It is to such patchiness and to such disturbances that many ecologists now look for an explanation of the fabulous species diversity of the Amazon.

The same is true of the deep-sea floor; the analogy that Sanders drew still holds, but its content is different. The deep sea is—as an oceanographer named Larry Madin once said in a different context—a big place made up of a lot of little places. Over great distances and over geologic time most of it is indeed a more stable environment than the land. But it is nonetheless susceptible to change on smaller scales of space and time. Disturbances still happen. Storms sweep through and trees fall (metaphorically but also literally), and in so doing they create new niches for life, just like the disturbances in the rain forest. Only lately, though, have man and his instruments been there to hear the trees falling in the deep. It has changed our conception of deep-sea life. What once was considered a desert is now more aptly seen as a mosaic of oases, and the tremendous and surprising diversity of abyssal life must have something to do with that simple fact.

To begin with—let us visit some of the places of the deep— the once-was desert contains large regions that for some time have been recognized as distinctive. The deep-sea trenches are a fascinating example. All around the Pacific they run, most of them plunging to depths of 30,000 feet or more, particularly in the western Pacific. But they do not form a continuous ring. They are interrupted by large gaps, regions in which the seafloor shoals to less than 20,000 feet or so—the depth deep-sea biologists conventionally regard as the beginning of a trench. If one could look at a bathymetric chart of the Pacific in photographic negative, or more precisely in topographic negative, the trenches would appear as chains of islands or long ridges, isolated from one another by large expanses of sea.

As far as the animal life in the trenches is concerned, islands are what they are. The trench fauna is still very poorly known. Most of that knowledge comes from the expedition of the *Galathea*, which trawled in five trenches in the western Pacific and Indian Oceans between 1950 and 1952, and from the Soviet research vessel *Vityaz*, which systematically explored most of the Pacific trenches between 1953 and 1962; since then, such research not being the sort that elicits eager government support in democratic countries, biologists have returned only occasionally to individual trenches. Both the Danish and the Soviet biologists found that the trenches were inhabited by a specialized group of animals—a "hadal" fauna, as it came to be called (although the freezing-cold muddiness in which these animals thrive is far from the popular conception of Hades). What is more, by and large each trench seemed to be populated by its own species of animal, just as one would expect to find on islands.

What lives in a trench? Mostly things that eat mud: bivalve mollusks (clams), polychaetes (bristle worms), and above all, holothurians, which is what biologists call sea cucumbers. "Without exaggeration it can be stated that the [hadal] is truly the kingdom of the holothurians," the Soviet biologist G. M. Belyaev wrote in summarizing the results of the *Vityaz* expeditions. Torben Wolff, a Danish biologist who sailed on the *Galathea*, went so far as to suggest that the fish Jacques Piccard saw through the porthole of the *Trieste* was not a fish at all, but a holothurian: after all, Wolff pointed out, Piccard was no zoologist, and flatfish have otherwise never been seen below 5,000 or 6,000 feet. Fish in general seem to be rare in the trenches, whereas holothurians, as both the *Galathea* and the *Vityaz* discovered, are present in large numbers.

They make unasssuming monarchs of the deep. Cylindrical and anywhere from one inch to six feet long—the ones in the trenches are at the short end of the scale—they are related to starfish, sea urchins, and sea lilies. All those animals are echinoderms, but the holothurians are the oddballs of the phylum.

Their skin is not spiny (which is what "echinoderm" means) but leathery soft, and the fivefold symmetry so characteristic of echinoderms—think of a starfish's five arms—is muted in them: it shows up as five rows of tube-feet that stretch from the animal's mouth at one end of the cucumber to its anus at the other. The feet are like protuberances on a water balloon. The animal has an internal hydraulic system of canals and reservoirs, and when it wants to extend any one of its dozens of feet, it simply contracts the appropriate muscles, forcing water into that foot. With these strange and slimy legs, holothurians crawl on the slimy bottom of the sea at a few body lengths per hour—although some, the weak-legged kind, do not so much crawl as plow through the top layer of sediment, leaving distinctive furrows. Others simply bury themselves mouth-first in the mud, with only their anus protruding into the water. Most holothurians breathe through their anus.

All of them must stop frequently to eat if they do not do so while underway. There is some evidence that holothurians can be picky eaters, selecting bits of sediment that are rich in organic matter, such as other animals' feces, from the great mass of mud. But if not, their long, coiled digestive tube does the selecting for them, extracting the food and excreting the rest in a long, coiled, and impressively large fecal cast. When pressed, holothurians also have the astonishing ability to excrete their entire digestive tract, and then to grow a new one over the next few weeks, just as their cousins the starfishes can regenerate any arms they might happen to lose in battle. The benefits to a holothurian of throwing out one set of guts and growing a new one are more obscure. One theory has it that the behavior is a way of alarming potential predators.

Another mystery about holothurians is their tendency to form herds. On several occasions the *Vityaz* brought up thousands of the animals, mostly of a single species, in a single trawl; it was these impressive numbers that led Belyaev to declare holothurians the kings of the trenches. But neither he nor anyone since has been able to say definitively why the animals tend to flock together. Possibly they do so when there is a good bit of food going; possibly it helps them find a mate in a lonely world. Once a herd of holothurian corpses is on the deck of a ship it is impossible to say what they were doing before they were caught. (That has always been one of the fundamental problems of deep-sea biology.) With the help of an automatic camera called Bathysnap, British researchers from the Institute of Oceanographic Sciences in Surrey have in recent years managed to photograph holothurian herds, thundering (so to speak) across the abyssal plain and leaving a trail of churned sediment in their wake. But they too have been unable to settle the matter of what the animals are up to.

In the trenches the holothurians and other animals are surprisingly well fixed for food. If anything, the supply is somewhat greater than in the neighboring, shallower regions of the abyss. That helps explain why the total mass of life in the trenches is also no smaller; the general rule that biomass decreases with depth breaks down at the lip of a trench. There are two reasons for the bountifulness of trenches. Most are close to land—they usually form where a seafloor-bearing plate collides with a land-bearing one—and thus to rich sources of food. The Galathea dredged coconut husks and bamboo shoots from the bottom of the Philippine trench. More important than the land itself, though, are the coastal surface waters, which tend to be much thicker with plankton than the open ocean is. Moreover, besides being close to food, trenches are by their very nature good at trapping it; as the organic detritus sinks from the surface waters, the trenches funnel it to animals on the bottom. But the animals pay a price for this service. Trenches are earthquake belts, and so from time to time a quake may jolt loose a great drift of sediment that has accumulated on the slope of a trench. The sediment slumps to the bottom in a turbulent avalanche, burying the animals and killing them.

In between such winnowing catastrophes, however, the trench animals enjoy the same temperatures as animals elsewhere in the abyss (around 36 degrees Fahrenheit), the same salinities (around 35 parts salt per thousand parts of water), and slightly more food. The one feature of their environment that is very different and more challenging is the pressure. It increases relentlessly with depth, at a rate of about 1 atmosphere per 10 meters, or 33 feet. At the bottom of the Mariana Trench the pressure is thus around 1,100 times what we feel when we stand on the beach. It is a pressure that would crush one of us in an instant, reducing the human body to a small fraction of its normal volume. (One of the favorite pastimes of young oceanographers on their first cruise is to lower Styrofoam cups—or better yet, Styrofoam heads from a hat store—to much more modest depths than those of the Mariana Trench, and then use the miniaturized objects to impress their landlubbing friends.) It is not known how the trench animals go about surviving the pressure they face. It seems reasonable to suppose, although some eminent deep-sea biologists have denied it, that they must have some special adaptations above and beyond the ones evolved by ordinary abyssal creatures.

In the case of bacteria, a biologist named Art Yayanos at the Scripps Institution has demonstrated as much. Yayanos has managed to bring back live bacteria from the bottom of the Mariana Trench and to culture them in his laboratory. One of the devices he uses is a block of titanium alloy, about the size of a large textbook, into which he has drilled an L-shaped chamber that holds about a pint of seawater. He puts a small piece of fish in the chamber as bait, and lowers the device over the side of the ship. At the seafloor a spring-loaded piston automatically closes off the entrance to the chamber, trapping whatever animal has been attracted to the bait, but trapping it at the pressure that prevails on the bottom. Once the trap has been hauled in, Yayanos can peer at the animal through tiny windows cut into the titanium block.

The animals he catches by this tactic are amphipods, small crustacean scavengers that resemble shrimp. (In fact they belong to a different class.) By keeping them pressurized, Yayanos has kept them alive at the surface for weeks. But eventually, after several weeks with no food, an amphipod dies and rots. The bacteria that do the rotting come, like the amphipod itself, from the bottom of the trench. By isolating them and culturing them in his laboratory, Yayanos has shown that bacteria from the Mariana Trench—and many other deep-sea bacteria—grow best at whatever exorbitant pressure they happen to be accustomed too.

Perhaps the amphipods and other trench animals have similar adaptations to specific pressures; and perhaps the range of pressures in the deep sea establish a range of ecological niches that different species of organisms can fill. "Ecologists have known that if you change the temperature and salinity a little bit, you change the life," says Yayanos. "But for a long time pressure hasn't been appreciated as a factor. You can't just pack your bags and move from a depth of 3,000 feet to 30,000 feet. It's not possible."

But the evidence that trench animals are specifically adapted to trench pressures is still slight. Aside from Jacques Piccard and Don Walsh, Art Yayanos and his lab mates, peering through minute peepholes into the high-pressure world inside their titanium block, are the only human beings ever to observe the behavior of a trench animal at first hand. The *Trieste* has long since been retired. And though Japanese researchers recently landed a tethered robot on the bottom of the Mariana Trench, no submarine around today can dive that deep.

There are other regions of the abyss, however, as insular as the trenches and yet very different in the circumstances they offer to life, that research submersibles have recently begun to explore. Between the trenches that rim the Pacific and the midocean ridges that bisect its eastern half lie vast realms that are far from featureless and whose topography consists of more than just wave after linear wave of abyssal hill. The Pacific abyss is dotted with isolated mountains called seamounts. They are extinct volcanoes, except for the few that are not extinct, such as Loihi Seamount, just off the big island of Hawaii. Some seamounts were created at a midocean ridge and have since been carried away by the motion of the plate on which they rest; others, like Loihi and the rest of the Hawaiian islands, were built by hot spots, isolated plumes of magma that rose from deep in the mantle and punched through the moving plate. Sonar maps being what they are, no one, except perhaps the U.S. Navy, knows just how many seamounts there are in the Pacific; estimates have ranged from ten thousand to a million. In the North Atlantic, according to data released by the Navy, there are 810 seamounts, but the Atlantic is much less active volcanically than the Pacific.

The exploration of seamounts as habitats rather than mere sonar blips has only begun in the past decade or so. They are in many respects the opposite of trenches. Rising thousands of feet off the abyssal plain, their summits may lie tens of thousands of feet above the bottoms of the trenches, and so the ambient pressure, while still pulverizing by human standards, is a fraction of what it is in Hades—which is why a submersible like the *Alvin*, out of Woods Hole, can dive onto them. At their high altitudes the peaks of the seamounts are also exposed to more intense currents, just as mountaintops on land are buffeted by winds. Yet seamounts lack the harshness and sterility of Everest; on the contrary, the life on them is, by deep-sea standards, downright lush.

The animals that do best near the tops of seamounts are the ones that live off currents rather than mud. In many places the slope of the seamount is too steep to allow mud to accumulate, or else the current has eroded the mud away, leaving bare rock. This would be tough terrain for a holothurian or any other detritus feeder; there is not much detritus to feed on. It is heaven, though, for a suspension feeder: an animal that is equipped to filter particles of organic matter from the passing water. The more water that passes—so long as the current is not fast enough to flatten the creature—the better off such an animal is.

Among the animals that make a living this way on sea-mounts are corals—not the reef-building kind, but ones that look more like exotic plants. In the mid-1980s Amatzia Genin and his colleagues at Scripps towed a camera-and instrument-laden sled, called Deep Tow, over Jasper Seamount, 350 miles southwest of San Diego. Jasper has a broad summit with at least six different peaks, two of which soar to depths of less than 2,000 feet. The peaks, Genin found, are where the action is. The Deep Tow photographs showed they were covered with fields of a spiral black coral, as many as 20 per square meter, clustered in eerie dark forests on every knob and pinnacle: on any perch that would allow them to intercept a decent current. Each spiral looked a bit like the dead branch of a corkscrew willow, planted in the hard rock. But each one was a colony of polyps, and each polyp was equipped with tentacles around its mouth to snare food from the passing water. Water hitting the slopes of the seamount, it seems, gets deflected up the slope, and as it rushes over the summit, it accelerates. This express-service food delivery is what the corals like; at mid-slope there were many fewer of them. A common feature of mountains on land is the tree line, above which no trees can grow; Jasper and other seamounts have a kind of inverse tree line.

But not all seamounts are the same. At Volcano 7, which lies around 400 miles southwest of Acapulco (and which deserves a better name), researchers have found a tree line that is much more sharply etched than the one at Jasper and that has a landlike orientation. Diving on Volcano 7 in *Alvin* a few years ago, Karen Wishner of the University of Rhode Island and her colleagues were surprised to find that the peak of the volcano, at a depth of 2,400 feet, looked virtually barren. A few small rat-tail fish swam about; the odd coral relative clung to a rock. But just 70 feet below the peak and down the slope from there, the situation changed dramatically. Wishner and her mates had dived into a crowd. There were rat-tails and crabs and shrimp and sponges and starfish, dozens of animals visible in a single glance out one of *Alvin*'s small portholes. There were large numbers of xenophyophores, single-cell, amoeba-like animals that grow as large as a human fist—they are the largest single-cell creatures on the planet—by gluing together minute plankton shells into ornate structures that look like honeycombs or lettuce. The abundance of life on the slopes of Volcano 7 was as startling as the barrenness of its summit. It was enough to gladden the heart of any deep-sea biologist.

And it seems to have a simple explanation. Not many things live on the summit of Volcano 7 for one of the same reasons not many things live on the summit of Everest: there is not enough oxygen. The eastern tropical Pacific is an oceanic backwater that lies outside the main ocean currents, and so the oxygen supply of its subsurface waters is refreshed only slowly. At the same time, its surface waters are rich in plankton, which are constantly dying and sinking and getting decomposed by bacteria—a process that uses up oxygen. The result is a thick layer of oxygen-poor water just below the surface that is about as dead as a polluted lake. The peak of Volcano 7 protrudes into the core of this layer, and so its population is sparse. Down the flanks of the mountain, though, the oxygen concentration increases to a level that more animals can survive, and—beautiful twist—those animals benefit from the poverty of the waters above them. The oxygen-poor layer is so oxygen-poor that a lot of plant matter from the surface sinks right through it without getting decomposed. Wishner and her colleagues saw mats of green plant matter lying untouched on the sea-floor. In other words, there is an unusual amount of food lying around on the sides of Volcano 7. That is why there is also an unusual number of animals.

Volcano 7 is probably a special case; most seamounts probably do not intersect an oxygen minimum. If seamounts in general are havens of abundant animal life, the reason in general must have something to do with the way they affect the currents and tides that sweep over and around them. The effect may be twofold. The intensified currents around seamounts may bring more food within reach of the outstretched tentacles of suspension feeders, as seems to be the case at Jasper Seamount. But the currents may bring more animals to a seamount as well as more food. Many deep-sea denizens reproduce by scattering scads of larvae in the open water, there to be dispersed by currents. For the larvae it is a risky business: they do not want to land on their parents, but they also do not want to settle in an unsuitable environment, one that will not allow them to metamorphose into successful adults. An animal adapted to the rocky pinnacles of a seamount, for example, would fare poorly on the flat mud of the neighboring abyssal plain. That is where the distinctive current pattern over seamounts may come in. A seamount seems to trap water in swirling eddies that remain lodged above the summit. It may thereby trap the larvae of the organisms that inhabit it, preventing them from being blown off their place of birth, their island sanctuary, into an unforgiving abyss.

Seamounts are obvious islands. (Sometimes, as in the case of Hawaii, they erupt so long and hard that they grow into actual islands, piling up lava until it breaks the surface of the sea.) Elsewhere in the ocean, along the long slopes that parallel the continents and across the vast abyssal plains, the island habitats are less obvious to the human eye; this is why a belief in the homogeneity of the deep sea could persist for so long. Yet the islands exist. It is just that they are very small, so small that they can only be seen up close—sometimes as small as the organisms themselves, because they are made by the organisms themselves.

The deep-sea floor was first seen by human eyes in 1947, when Bruce Heezen, using an automatic camera developed by his teacher Maurice Ewing, took photographs of the continental slope off New England. Throughout a career spent studying geology on a global scale, Heezen remained fascinated with deep-sea photographs and their up-close view of the seafloor. In 1971 he and one of his own former students, Charles Hollister, now at Woods Hole, published a photographic tour of the seafloor, which they called *The Face of the Deep*. It was a marvelous and idiosyncratic tour, and it is still one of the best sources of portraits of abyssal animals. Heezen and Hollister had combed through a hundred thousand photographs, photographs of a sparsely inhabited environment that had perforce been taken blindly and more or less at random, to find the best shots of life. At one point they let the reader in on what most of the rejects had shown. "Mounds, mounds, mounds, mounds," they explained, "that is about all one sees in the average ocean-floor photographs . . . just mounds, mounds, mounds, mounds."

The mounds are made by animals. In the Santa Catalina Basin off southern California, at a depth of around 4,000 feet, mounds cover two percent of the seafloor. They are typically four inches high and a foot across, and they are made, most of them, by a large echiuran worm. Echiurans are burrowers, and their burrows are often U-shaped and two-holed. The buried worm itself is short and fat. But it has an extraordinarily long proboscis, more of a tongue really, which it extends out one of the holes. Flicking slowly in and out, the tongue licks up sediment in a circle around the hole, sometimes forming a lovely rosette pattern in the soft mud. Never leaving their burrows, the worms in the Santa Catalina Basin dig shallow pits about a yard across with their tongues (which must therefore be at least a foot and a half long). The hole at the other end of the U is where the worm deposits its excrement, so as not to be poisoned by it. In periodic eruptions, the animal slowly builds a volcanic mound of fecal pellets on the seafloor.

Actually the process is not as slow as one might think. Craig Smith of the University of Hawaii and Peter Jumars of the University of Washington have closely observed the Santa Catalina echiurans, and one unsuspecting worm in particular. Diving in *Alvin*, Smith and Jumars placed a camera tripod directly over the worm's fecal mound and set the camera to take stereo photographs of the mound every ten hours. A month and a half later they returned in *Alvin* and collected the camera. The photographs showed that during the first five days of the experiment, the worm had added half an inch or more of excrement to the top of its foot-wide mound. Then it had spent the next month resting. Other experiments done by Smith and Jumars suggested this was typical echiuran behavior and not just the aberrance of a worm terrified by strobe lights.

Strange as it may seem, the echiuran's behavior has important consequences. In most of the deep sea, sediment accumulates at about the rate dust settles on your furniture; so if a worm deposits half an inch of sediment on its mound in a few days, that is an extremely rapid rate. Smith and Jumars did a simple calculation, based on the number of worm mounds in the Santa Catalina Basin and the rate at which they are built. They found that echiuran worms eat and excrete the top four inches of sediment (the depth of their feeding pits) in the entire basin every 70 years. That means no grain of sediment gets permanently buried, escaping biology to become geology, without going through a worm gut around ten times. And the Santa Catalina Basin is not an especially worm-rich area; it is reasonably typical of the deep-sea floor. As Heezen and Hollister attested, mounds of the kind found off southern California are ubiquitous in the deep. The ocean floor, then, is constantly being churned by large worms and other animals.

That cannot fail to have an effect on the smaller animals that live in the mud—the infauna, as biologists call them. Smith and Jumars built their own mounds, ones that were devoid of life, from sediment they had removed from the floor of the Santa Catalina Basin. During the same *Alvin* dive in which they deployed the camera tripod, they placed their artificial mounds on the seafloor. When they came back a month and a half later, they found that the mounds were crawling with minute polychaetes, segmented and bristly like caterpillars. To these worms the mounds were virgin territory or fresh food or something—their kind of island, in any case. Apparently the polychaetes are adapted to colonizing the mounds that are constantly being erected by their larger brothers, the echiurans.

The islands that deep-sea organisms colonize need not be made of mud or excrement. They may also consist of other living animals. Xenophyophores, the strange single-cell animals that are abundant on Pacific seamounts and just about everywhere else in the abyss, are a good example. When they were discovered by the *Challenger*, they were mistaken for sponges, and for most of the following century they were either ignored by biologists or mistaken for mud. And indeed, by the time they are on deck, their fragile and elaborate tests—shell-like structures—have been crumbled by their rough treatment in the dredge, and mud is about what they look like. But with submersibles and deep-sea cameras it has become possible to see these weird creatures for what they are—not mud, and not just single organisms, but whole colonies of life.

The living part of the xenophyophore itself is a single, giant cell that is contorted and branched and enclosed in a filigree of tubes. Somehow and for some reason the animal draws mineral crystals into the cell plasma (*xenophyophore* is Greek for "bearer of foreign bodies"); and somehow the cell surrounds itself with a test assembled from sand-grain-sized shells of dead plankton. The test is studded with fecal pellets, which the animal apparently considers worth saving; they may be colonized by bacteria that recycle the feces back into food for the xenophyophore. The ornate test itself helps the animal trap fresh food, which it probably draws into the cell with the help of pseudopods that extend from the tubes. Otherwise the animal is immobile and even rooted in the mud.

There it is colonized by far more than bacteria. A xenophyophore is a single-cell organism that, in a reversal of the usual order of things, serves as a home for animals that are far more complex. On Volcano 7 in the eastern Pacific, Lisa Levin of North Carolina State University found one xenophyophore that was harboring a hundred animals in and around its five-inch wide test. Thirty brittle stars, a common starfish relative with long, thin arms, crawled over the test or loitered under it, hiding, perhaps, from predators. Seventeen polychaetes sat in the xenophyophore's various crannies, waiting, perhaps, for food to be trapped from the water and funneled their way. Thirteen isopod crustaceans were also present, and among them were males, females, and young; this suggests that the isopods had taken up permanent residence and were breeding. Shelter, food, and procreation are what one mainly looks for in a home, and xenophyophores seem to offer all those things. They are common, too: Levin found more than two per square meter on the flanks of Volcano 7. On the abyssal plain off Mauretania, West Africa, British researchers have photographed fields of xenophyophores that are ten times as dense. They look like sparsely planted lettuce patches. Wonderfully enough, these single-cell homemakers often live on the sides of mounds built by other animals—colonies piled on colonies.

The mound builders and the xenophyophores are two examples among many of a general principle of the seafloor: even where it appears flat, the animals that inhabit it create their own topography, their own relief at their own scale, their own plethora of niches. From homogeneity they create spatial heterogeneity. From a uniform sea of mud they create islands of habitable space, from a featureless fabric, a patchwork quilt. But in this great project they have help. Patches on the seafloor do not all have to be created on the seafloor; there is another way. They can also fall from above, like manna.

Until the late 1960s, when John Isaacs, a college dropout cum commercial fisherman cum engineer cum marine biologist at Scripps, had the idea and developed the technology for lowering a camera with bait attached to it into the deep sea, the general notion was that the seafloor was inhabited almost exclusively by the sort of docile invertebrates that we have just been talking about—echinoderms, worms, corals, and the like. And in fact, in terms of sheer numbers, those animals do dominate the deep. But Isaacs and a succession of Scripps or Scripps-trained biologists have shown that lethargic crawlers, burrowers, and sitters are not alone on the ocean floor.

When you lower some nice, fresh mackerel, say, into the abyss, you attract the active interest of all sorts of animals that are generally smart enough and fast enough to keep their distance from a dredge. Isaacs called his first device the Monster Camera. The monsters he and his followers photographed were animals that were adapted to a different kind of patch than an echiuran mound or a xenophyophore. They were adapted to a patchy food supply, one that arrived unpredictably from above, at random times and places.

When Bob Hessler and Art Yayanos lowered a Monster Camera into the Philippine Trench, for instance, they did not find it to be the kingdom of holothurians: instead it seemed to be ruled by tiny scavenging crustaceans, amphipods of the family Lysianassidae (the same amphipods that Yayanos later used as bacteria farms). The first inch-long amphipod arrived at the bait within minutes, and within less than an hour the bait was concealed by a frenzied swarm of animals too numerous to count. Elsewhere in the deep, biologists encounter much the same scene: they lower bait, and a short time later the first amphipod arrives, beating upcurrent toward the bait, which it can apparently smell, and swimming as fast as five body lengths a second—a speed comparable to that of the fastest human sprinter. The animal stays at the bait for half an hour or so, during which time it may consume half its body weight in mackerel. Then it leaves, apparently sated. Depending on how many of its mates have joined in the frenzy, the bait may be reduced to scraps in as little as a couple of hours.

Tiny crustaceans are not the only thing attracted by baited cameras. Big fish are too. Among the most common are rat-tails, so named because of their long, slender tails (some of which are more graceful and less ratty than others). Rat-tails of the genus *Coryphaenoides*, also called grenadier fish, get quite large—more than six feet long, judging from photographs. Ken Smith of Scripps and Imants Priede of the University of Aberdeen have photographed grenadiers in the Pacific and the Atlantic. Smith used to think that grenadier fish just sat and waited for food to fall near them; if food is scarce, he reasoned, an animal cannot afford to waste time and energy searching for it. But then Smith figured out a way to track grenadier fish after they left the camera's small field of view. He and Priede embedded small sound transmitters, the size of a little finger, in the bait. Grenadiers ate the transmitters and then swam away, their stomachs emitting regular ultrasound pings that were picked up by a hydrophone on the camera tripod, as long as the fish were no more than 500 yards away.

As it turned out, the fish did not stay close for very long. Like amphipods, grenadiers arrive at a baited camera swimming upstream, following the smell of the mackerel, and they leave in fairly short order—within an hour, in one of the places

Smith and Priede dropped their camera. From the time it took the first grenadier to reach the bait at that North Pacific station (16 minutes), Smith and Priede calculated that, on average, an acre of seafloor there must be populated by between two and three grenadiers. That would give each fish about the same territory as a typical suburban homeowner. But grenadiers do not seem to be territorial; nor do they simply drift on the current waiting for deep-sea biologists to drop a mackerel their way. Instead they are constantly roaming, searching for food. They stick around the bait only as long as it makes sense to, given the likelihood in that region of finding more food elsewhere; in regions where there is a lot of food, they move on quicker—a strategy biologists call "optimal foraging." At a stately pace of a quarter of a mile per hour or so, but never stopping, a grenadier may log as many as 2,000 miles a year.

The reason this strategy makes sense for grenadier fish is that it is not only biologists who drop patches of food onto the ocean floor. Nature's food fall is patchy too. Cruising around the Santa Catalina Basin in *Alvin*, Craig Smith (no relation to Ken) has seen a dozen or more animal carcasses, some of them covered with scavengers. He has even seen a dead whale. Some years ago Smith put out baited cameras to investigate who exactly the scavengers in the Santa Catalina Basin are. There turned out to be a lot of them. First to arrive at the bait, within minutes, were primitive hagfish, blind, jawless, and eel-like. Forming a writhing Gorgon's head around the mackerel, they drilled into it and consumed it from the inside out, all the while secreting clouds of mucus to deter competition from other scavengers. The amphipods, though, were not deterred: they swarmed on through, eating the hagfish mucus and some of the bait as well. Next came brittle stars, thousands of them, piled several deep in a tangle of arms, waiting shyly for hours or even days for the hagfish to disperse; then they would surge en masse over the remaining scraps. Even 19 days after Smith deployed one of his cameras, the excitement it caused was not quite through: a lone crab and a sea cucumber were still picking over the mackerel bones.

The hagfish got 90 percent of the mackerel in Smith's experiment, and for them it may have represented a substantial fraction, if not all, of their annual diet. Smith saw several hagfish with bulging guts lying motionless on the seafloor for several days, apparently digesting their meal. The other scavengers probably find many sources of food, only some of which are likely to be fish carcasses. A dead fish is a comparatively rare windfall in the deep sea. If the only thing patchy about the seafloor were its supply of dead fish, it would not be a very patchy environment at all.

What deep-sea biologists have come to realize in the past two decades, however, is that much of the food available to animals on the ocean floor comes in patches. Most of it is much smaller than a dead fish. It consists of microscopic corpses: the remains of the single-cell plants that float at the surface of the sea, called phytoplankton, and of the single-cell animals, called zooplankton. Fecal pellets from the zooplankton are thrown into the mix as well. All these various gooey particles stick together to form large, visible flakes that are heavy enough to fall through the water. The flakes are known as marine snow. Every diver has seen it; it is everywhere in the sea; it seems a constant. And not long ago, that was the image biologists had of the deep sea: it was a world where the snow and thus the food fell softly, thinly but constantly, like a never-ending winter night.

It is true that the snowfall never ends. But it is not true that it is constant, and in retrospect it is hard to see why anyone ever thought it would be. After all, the supply of phytoplankton living at the sea surface is not constant. There are seasons in the sea as there are on land. The plankton bloom in spring, and as spring gives way to summer they die in large numbers. Then the light snow that is always falling through the deep becomes a blizzard, and great pulses of dead plankton—phytodetritus, oceanographers call them—sink through the water. The first evidence for such seasonal pulses of sediment was collected in 1980 by Woods Hole oceanographers who used an exceedingly simple tool: they hung vertical, conical sediment traps, the equivalent of rain gauges, in the water column at a depth of 10,000 feet off Bermuda. Since that eye-opening discovery, other researchers have seen the pulses, with their own eyes and in photographs, in many parts of the ocean.

The dead matter sinks so rapidly that it arrives at the seafloor green and fresh. Wishner and Levin saw green flocculent mats on the flanks of Volcano 7 in the eastern Pacific. Southwest of Ireland, in 1982, Anthony Rice and his colleagues at the British Institute of Oceanographic Sciences made time-lapse photographs that showed pulses of green detritus arriving at the seafloor in a region called the Porcupine Seabight. The first pulse arrived in late June and spread out to form a small patch. Soon thereafter the pace quickened, as patch after patch rained in, covering the brown mud with overlapping tiles of dark green. The snow pooled in the furrows and pits dug by the seafloor animals, and it drifted up against the side of their mounds. Sometimes a passing gust of current caused it to billow cloudlike back up into the water, obscuring the camera's view, or to roll across the seafloor in a loose ball, like tumbleweed. By late July, the mud around the camera was fully carpeted in green.

But by August the green was starting to disappear again. Seafloor animals, it seems, respond rapidly to the blizzard of food. Rice and his colleagues have photographed holothurians grazing placidly on a patch of fresh phytodetritus. They have seen the same brittle stars that Craig Smith saw eating mackerel in the Santa Catalina Basin eating phytodetritus in the Porcupine Seabight. Smaller and less conspicuous animals respond to a new food patch as well. Single-cell foraminifera may swarm over it, eating the plant matter and then reproducing wildly, in a population explosion that mirrors the bloom at the sea surface.

In the Norwegian Sea, a German biologist named Gerd Graf has found that tiny, burrowing worms called sipunculids manage to snatch away much of a food fall before forams or other animals can get it. Graf hauled up undisturbed blocks of sediment with a box corer, a device that Jumars and Hessler had developed in the 1970s. He found that the sediment was riddled with worm burrows, around a thousand of them per square foot, with an individual worm having around 20 burrows of its own. At a depth of a few inches in the mud, those 20 burrows merged into a single tunnel that extended to a depth of nearly two feet. The worms apparently lie waiting in their burrows for their annual or biannual pulse of food (a smaller plankton bloom occurs in the fall). When it comes, they stick their heads out of their burrows and haul the loot in within days. They may squirrel it away somehow for leaner seasons.

More than a century ago, when Moseley of the *Challenger* was describing the monotony of life in the deep, and helping to produce a mind-set that would endure until the 1960s, he nevertheless allowed the possibility of a "periodical variation in the supply of food falling from above, which may give rise to a little annual excitement amongst the inhabitants." He could not suspect how right he would turn out to be in that one offhand remark, or how wrong he was in general.

———

J. Frederick Grassle has devoted most of his career to trying to figure out why the deep sea is not at all the way Moseley described it, and why so many different things live there. Grassle is now the director of the Institute of Marine and Coastal Sciences at Rutgers University, but he got his start and spent most of his career at Woods Hole. In 1972 he took Bob Hessler's place working with Howard Sanders. (Hessler had moved to Scripps.) In the 1980s, Grassle went on to produce more accurate estimates of deep-sea diversity than Hessler and Sanders had done. Instead of an epibenthic sled he used a box corer. For all its virtues, the epibenthic sled still missed a lot of burrowing organisms; worse, it jumbled together animals from an area of the seafloor whose size was unknown—the sled was dragged for somewhere around a mile—which made it impossible to be precise about just how many species there were in a given place. The box corer allowed Grassle to be more precise: it allowed him to count everything that lived on and in one square foot of ocean-bottom mud.

The numbers of species he got were even more startling—that is, higher—than the ones Hessler and Sanders had come up with. Grassle and his colleague Nancy Maciolek took 233 box cores along a 110-mile-long section of the continental slope off New Jersey, mostly at a depth of 7,000 feet. They counted a total of 798 species: 798 different species of animal living in 233 square feet, which is the size of a not especially palatial living room. But more significant even than the raw number was the pattern of the diversity. As Grassle and Maciolek progressed along the 7,000-foot-depth contour, the number of species rose steadily, with no sign of leveling off. Each new box core could be counted on to bring in new species. In other words, Grassle and Maciolek were nowhere near having sampled all the things that lived in that particular region of the deep, at that particular depth; 798 was just where they stopped counting.

Extrapolating from their experience and trying to be conservative, the two researchers attempted to estimate what they would find if they were endowed with a godlike ability to number all the species of animals living on all the ocean floor. Quantifying ignorance is an inherently uncertain thing to do, but no one was better placed than Grassle and Maciolek to have a go at it. "As more of the deep sea is sampled," they concluded, "the number of species will certainly be greater than 1 million and may exceed 10 million." These are numbers that place the deep sea on a par with the rain forest as a haven of diversity. (And if the rain forest demonstrates, in J. B. S. Haldane's famous dictum, that God has an inordinate fondness for beetles, then the seafloor proves His predilection for polychaetes: nearly half of the species collected by Grassle and Maciolek were worms belonging to that class, and more than half of those were new to science.)

After twenty years of reflecting on why the seafloor should have such a fundamental similarity to the rain forest when it looks so fundamentally dissimilar, Grassle has arrived at a theory that is somewhat different from that of his early mentor Howard Sanders. "There's more than one reason for the diversity," he says. "But I think probably the strongest argument is that it's a bit like the formation of gaps in the rain forest—like trees falling down. There you get an increase in nutrients that favors some species, and it's this mosaic of small patches that helps maintain the diversity. The way that mosaic occurs in the deep sea is from food coming down from the surface. It could be a patch of Sargassum seaweed, it could be a piece of wood, or it could be a patch of phytoplankton accumulating in an animal burrow or a natural depression in the seafloor. That organic material has a very uneven distribution. And it's that mosaic of different amounts of food which seems to be very important for maintaining diversity—each of these sources of organic material creates a mosaic that allows a lot of species to coexist.

"The reason is, when you open up a resource of food, the rate of response is different for different organisms. And the probability that two species get there at the same time is low, just because the food is patchy and widely separated. So having patches is a mechanism for reducing the possibility that species will compete with one another, with one species eliminating another. If an environment is absolutely constant and unchanging, finally one species wins out. The more you can do to make the environment both spatially and temporally complex, the more species can coexist."

There are other ways of creating complexity, Grassle says, besides dropping intermittent patches of food. A tree falling in the forest is a food patch, because in rotting it replenishes the soil with nutrients. But it is also a disturbance, one that kills or drives off some animals and makes room for others. Disturbances promote diversity by providing opportunities for opportunists—the pioneer species that specialize in colonizing virgin land—and by thus keeping a habitat from being taken over by a few dominant species. A clearing in the forest, for instance, allows meadow grasses to live next to ancient trees and helps prevent the whole region from becoming an unbroken carpet of fir or pine. In this sense, too, there are always trees falling in the deep. The mound builders in the Santa Catalina Basin are always creating disturbances; the fresh mounds are rapidly colonized by polychaetes. A dead fish on the floor of the Santa Catalina Basin is also a disturbance. As the frenzied hagfish thrash and tear at the fish and as the brittle stars stampede toward it, the patch of sediment around the fish is roiled and flattened. According to Craig Smith, one of the most common polychaetes in the basin, a species that builds fragile balls of mud and lives in them, is either killed or driven out of that patch. Its homes are destroyed by the feeding frenzy—but two species of less common and more opportunistic polychaete colonize the churned sediment and thrive in it. Disturbances are equalizers that allow the underdogs of evolution to survive.

If a disturbance happens over a large area, on the other hand, it reduces complexity and diversity instead of increasing them. When a huge tract of rain forest burns down, all the multifariousness of that tract gives way, at least for a time, to a few grasses and other colonists (or crops in the case of a fire set intentionally by a farmer). If species diversity is like the many hues of a patchwork quilt, then a large disturbance is like covering most of the quilt with one giant patch of a single color. Such disturbances happen in the deep sea. The earthquake-triggered avalanches on the slopes of deep-sea trenches are one example, and the turbidity currents that roar down canyons on the continental slope are probably another.

A third example, which was discovered in the late 1970s by Charles Hollister, among others, are abyssal storms. On the continental slope south of Newfoundland, and presumably in other places that have not been studied, strong abyssal currents occasionally spin off large eddies—analogous to hurricanes in the atmosphere—that sweep along the slope and roil the sediment for days. Huge clouds of murk are kicked up into the water, and when they settle on the seafloor they transform it into something close to a clean slate. Like the trench avalanches, they wipe clean the fragile topography the animals had so carefully constructed, the mounds and burrows and feeding pits. Some animals thrive under these conditions. But both in trenches and along the tracks of abyssal storms, the overall species diversity seems to be reduced.

Yet such large disturbances appear to be rare in the deep. In most areas, animal mounds and animal tracks survive for a long time, even long after their creators have departed, which is why the seafloor is so covered with them even though it is sparsely inhabited. In that sense Howard Sanders was right: the fact that, on a large scale, the deep is comparatively stable over time—compared with shallow and more storm-tossed waters—is a prerequisite for explaining why it is so diverse. The fact that the deep sea is large, and on that scale spatially uniform, is also important: the largeness allows room for many different patches, and the uniformity means that there are no barriers that prevent organisms from dispersing over long distances and colonizing new patches as they become available.

But if Grassle is right, this stability and this uniformity are to the incredible richness of the deep as canvas is to a painting.

Without the canvas there would be no painting, yet it is not the canvas that makes the painting rich. And it is not the constancy of the deep but the constant change that makes it so rich in animal species. It is the constantly changing supply of food, raining down not in a steady drizzle but in periodic cloud bursts; the constant restless activity of the animals themselves, always reshaping their own environment and creating new nooks and niches for others; the constant small disturbances of the peace that allow different species to coexist peacefully. All these small changes, against a backdrop of stability, create an intricately textured environment in which all manner of living thing can find a welcoming home. "We're brought up to believe that the abyss and the hadal zone are really hostile environments to all life, because they're hostile to us," says Grassle. "But in fact, for a lot of life they're not hostile. It may in general be a more 'benign' environment, to use the word Howard Sanders used, than most parts of the planet. And that's despite the fact that it's cold, and under high pressure, and not a lot of food gets down there."

In February 1977, before these ideas about deep-sea diversity had crystallized in his mind, Fred Grassle published a paper in *Nature*, the prestigious British science journal, describing some experiments he had done at a deep-ocean site in the western North Atlantic. The paper was called "Slow recolonisation of deep-sea sediment." Grassle had dredged sediment from the site, allowed it to warm up, and then frozen it until he was sure everything in it was dead. Then, with the help of *Alvin*, he had replaced it on the ocean floor in shallow boxes.

Two years later he had come back to see what kind of life had moved into his boxes. The answer was not much: some worms, a few clams, some isopod crustaceans, but the population density in the boxes was still nearly a hundred times less than on the surrounding seafloor. It was a thousand times less than what Grassle found in a box he placed in the 30-foot depths of Buzzards Bay, right off Woods Hole. "Life processes as reflected in rates of colonisation and growth are remarkably slow in the deep sea," Grassle concluded in his paper.

There was other evidence for that conclusion, including some from an especially strange and unintentional experiment. In 1968, *Alvin* had sunk in 5,000 feet of water, when a cable snapped as the submersible was being lowered into the water from its mother ship. Just before the sub disappeared beneath the waves, as water poured in through the open hatch, the three men who were already inside, ready to dive, had managed to scramble to safety. But they had left their bag lunches behind. Those lunches, three apples and three baloney sandwiches, sat on the ocean floor for ten months, until *Alvin* was finally found and recovered. When the lunches were pulled out of the bag, they looked good enough to eat. Some people were brave enough to try them; they tasted salty but fine. They had not rotted at all. Life processes, as reflected in the rate at which bacteria degrade food, seemed to be much slower in the deep sea than in a typical refrigerator.

And just a couple of years before his 1977 paper, Grassle himself had participated in an equally influential experiment that had confirmed the lazy pace of life in the deep. He and Sanders had picked through garbage cans full of mud collected by the epibenthic sled and had pulled out numerous specimens of a deep-sea clam called *Tindaria callistiformis*. They had sent the specimens, which were around a third of an inch long, to a laboratory at Yale University that was able to do radioactive dating. From the amount of radium-228 in the clam shells, the Yale researchers estimated the age of the clams: They were around 100 years old. A century to grow a third of an inch is slow indeed.

The slowness of most deep-sea life has been confirmed again and again—but the most breathtaking exception to that general rule was being discovered just as Grassle was publishing his *Nature* paper. In February 1977, *Alvin* was beginning a series of dives onto the crest of the Galápagos Rift. The dives were taking place at the very site where, five years earlier, Ken Macdonald had heard the rumble of earthquakes and had seen dead fish floating at the surface. They were taking place at a site where, in the years that followed Macdonald's measurements, Scripps geologists using Deep Tow had discovered water that was slightly warmer than the surrounding ocean, as well as chemically distinct. In 1976, Deep Tow had also photographed clusters of strange, white, and unusually large-looking clams on the ridge crest.

The geologists were not interested in clams; one of them suggested the clams were probably just empty shells that had been dropped overboard by a passing ship after a clambake. But geologists everywhere were interested, a decade after the triumph of plate tectonics, in finding the first direct evidence of volcanism on a midocean ridge. (Macdonald's earthquakes were indirect evidence.) So in 1977 a group of them from Woods Hole, MIT, and Oregon State University took *Alvin* to the Galápagos. Using a map prepared by Scripps graduate student Kathy Crane, they went looking for hot springs on the ocean floor. Crane had never bought the line that the giant clams she had seen on Deep Tow photographs were garbage; she thought the clams were clustered around the hot springs that everyone was hoping to discover. Before Jack Corliss of Oregon State and John Edmond of MIT got into *Alvin* for the first dive, Crane recalls, she told Corliss just to follow the bright white clams when he got to the seafloor. They would probably lead him to what he was looking for.

And so it was that in February 1977, as Grassle's paper reflecting the old consensus on deep-sea life was appearing in print, two geologists were getting a close-up look, through *Alvin*'s portholes and the shimmering-hot water, at a new biological reality: at giant tube worms that grow in thick clumps, with blood-red heads like flower buds; at six-inch mussels and blind, porcelain-white crabs; at foot-long clams that grow, not slowly, but an inch and a half a year; at all this life thrown together in dense piles whose lushness rivaled that of any eco-system ever seen before, on land or sea, shallow or deep. Deep-sea biologists today still remember where they were when they heard the news of the discovery; among them it resonated like a moon landing. Before *Alvin* returned to port, Grassle and his colleagues were planning their own trip to the Galápagos, to the most spectacular patch of all.

CHAPTER 5

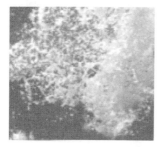

Springtime

Ken Macdonald did not have a submarine with him when his sonobuoys detected earthquakes off the Galápagos in 1972. He could not dive down to the seafloor to see what was going on. What might he have seen if he did? In April 1991 Rachel Haymon, a marine geologist at the University of California at Santa Barbara who happens to be Macdonald's wife, got an idea of what he had missed two decades earlier. Haymon was diving in *Alvin* a thousand miles or so northwest of the Galápagos, into the troughlike volcanic caldera on the crest of the East Pacific Rise—latitude 9 degrees, 50 minutes north; longitude, 104 degrees, 18 minutes west; depth, 8,500 feet below the ocean surface. It was a place she thought she knew well, a place where in 1989 she had seen hot springs teeming with animal life. But it was also a place where, just three years earlier, Macdonald and Jeff Fox had predicted an eruption might happen.

This is what Haymon saw: "The whole area had completely changed. In places that had been loaded with animals, we could drive back and forth and find only these little tiny crabs that looked like they were five days old. We drove all over and we didn't find any dense communities of hot-spring animals— zero tube worms. We would find hydrothermal vents in the general area we expected, but everything looked different."

Dan Fornari, a marine geologist at Lamont-Doherty Geological Observatory, was in charge of the cruise along with Haymon: "The first thing a marine geologist thinks about is 'Are you lost? Have you screwed up in some fundamental way?' But after all this gnashing of teeth, we finally figured out, 'No, we're in the right place. We know that.' Then the sub went down again, and the divers couldn't see a thing. I mean it was inky black—it was this smoky water that you just could not see anything through. On that dive they didn't do much more than drive around in the smoke. Well, this was not what it had been like in 1989. This was a dramatic change."

Haymon: "Then on our 14th dive, we found this community of thousands of tube worms and mussels—dead. They were scorched and shredded and blown up. Some of them were still alive, looking absolutely pitiful. There were some mussels that were opened up, and the meat was just lying there—nothing had touched it. There were hardly any scavengers, like crabs, which given all this free food, all these dead and dying animals, was really amazing. And on top of all this was what looked like a gray ash. It had pyrrhotite and sulfate—hydrothermal minerals—in it, and pieces of dead flesh. Whatever happened was a sudden violent event that destroyed the animals and rained all this ash on top of them. We collected some of the dead tube worms, and when we brought them up on deck the smell was kind of like hamburger on the grill."

Richard Lutz, deep-sea biologist, Rutgers University: "One of the tube worms that came back had what looked like a chunk of black basalt at the bottom—tube worms are sessile, and they're often attached to basalt—but when we looked in detail it was not basalt, it was charcoal-broiled trophosome, a piece of the tube worm's tissue. That was one reason for calling this site the Tube-Worm Barbecue."

Karen Von Damm, water chemist, University of New Hampshire: "The most impressive thing was there was hot water coming out everywhere on the sea floor. Usually you see it coming out of a few cracks. Here it was everywhere, and it had this white stuff in it. Looking at the pictures now you don't get a feel for the areal extent of it. It was literally like driving through a blizzard—not a snowstorm, a blizzard. The sea floor on a ridge is usually black, but here it was all white. It was amazing!"

Haymon: "What it was was bacteria—white, fluffy material, that was getting blasted up into the water, 150 feet above the seafloor, by hydrothermal venting. We put out a marker, and two weeks later it was just covered by this stuff."

Fornari: "The flocs of bacterial matter were being blown out of holes, of little skylights in the volcanic rock—we ended up calling them snow blowers, because that's exactly what it looked like. In some places, the whole caldera was this ghostly chasm of fresh volcanic rock, black and glassy, covered in white bacterial mat with these snow blowers going. It was an astounding, astounding terrain."

Von Damm: "We saw one vent that was 389 degrees Celsius [732 Fahrenheit] one day, then 396 degrees a week later, and 403 degrees a week after that. This was 400-degree water, and it was pouring out of a pile of rock! Basically it had boiled under the sea floor. You can think of it as steam, but it's under such high pressure that it's like a liquid. We thought boiling water might exist on the sea floor, but we had never found it."

Fornari: "So we knew that something was up. It didn't dawn on us right away because we were unwilling, I think, to accept the obvious—no one had ever witnessed an eruption on the sea floor before."

Haymon: "We gave some of the lavas to Ken Rubin at Scripps, who has a technique for dating young lavas. He calculated that the eruption happened between March 28 and April 6. Our first dive was on April 1. So the time around when we were there was when it happened.

"What I believe happened is that magma moved under the surface and displaced hot water. The magma is coming up from below, and it's like a piston, and it has to displace the water in the void spaces in the crust. But the boiling water's being confined from above by a thin cap of magma that had flowed out over the arca earlier and cooled and hardened. So the pressure built up until the cap was just blown away. We saw that the cap was pretty thin, because it had buried the tube worms only partially. It could have been a matter of only a few minutes between the flow that formed the cap and the explosion—lava cools immediately in cold seawater.

"Then you have water rushing out of the sea floor, flying tube worms, and glass flying all over the place. The glass may be what tore up the tube worms—or maybe they just exploded from internal heating, like hot dogs in a microwave. The hydro-thermal minerals that formed under the sea floor, I think, blew out with the water, and formed the gray ash we saw."

"So the excitement of this area for us is that here we have a time zero. We can mark the system, and see how fast the black-smoker chimneys will grow from minerals precipitating out of the vent smoke, how fast the animals will return, whether we'll continue to have eruptions or not. We'll start to have a feeling for how the system changes through time. We know that any one segment of the ridge will erupt sooner or later. But to catch it and actually track it has proved to be very difficult. We were just very lucky."

Lutz: "I was the only biologist on a cruise of geologists. Whenever they brought up something slimy, they would yell 'Where's Lutz?' I was like a pig in mud."

Haymon: "Everybody was jazzed."

Von Damm: "I've been going to sea and down in the sub for a long time. It used to be that people said one of the last great things to see on the seafloor would be an eruption actually happening—to see the red lava flowing on the seafloor. We don't really know what it would be like. It might be a very turbulent event. But after coming this close, I no longer have that desire. I don't want to see red lava flowing on the seafloor. A few days later is close enough for me."

Fred Grassle and the other biologists arrived at the Galápagos hot springs seven years later: seven years after the eruption that gave birth to them, assuming it coincided with the earthquakes and the fish kill that Macdonald witnessed in 1972. Until Grassle's cruise in 1979, deep-sea biologists had devoted very little effort to studying life on midocean ridges. Indeed, they had deliberately avoided the ridges, which, as Hessler once put it, were "graveyards for their gear." An epibenthic sled or a dredge or a trawl likes to be towed over flat seafloor covered with soft mud; in rugged mountain terrain it is apt to get snagged, torn from its cable, and lost. Deep-sea biologists have never enjoyed the sort of government largesse that would allow them to be cavalier about the loss of a major piece of equipment.

Nor was there any reason for them to suspect some spectacular new life on the crest of a ridge. Geologists had predicted that hot springs would exist on the ocean floor—it stood to reason, given the presence of volcanoes—and had gone looking for them. Only one biologist seems to have shown any degree of prescience, and only a slight degree at that, concerning what the geologists would discover on the Galápagos Rift. In his book *The Ecological Theater and the Evolutionary Play*, published in 1965, the famous Yale ecologist G. Evelyn Hutchinson had remarked that it might in theory be possible for organisms to make a living off the internal heat of the Earth, rather than off sunlight. But Hutchinson had no idea how the animals might do it, and he certainly did not have the fabulous communities around seafloor hot springs in mind. No biologist did have or could have. Nothing could have prepared a deep-sea biologist accustomed to dredging worms and such like from the sparsely settled abyssal plains and continental slopes for the sight he would see on diving onto the Galápagos Rift in 1979. It was like being born and raised in Labrador, in complete ignorance of the outside world, and then one day parachuting into Times Square.

In most of the deep sea life is sparse; at the Galápagos hot springs, there were places where the living matter was packed more densely than in a patch of cornfield the same size. In most of the deep sea the organisms are small; at the hot springs, there were tube worms eight feet tall—they grow upright and are rooted to the spot—and clams as much as a foot across. Grassle gave one such clam to the same Yale researchers, led by Karl Turekian, who through radiometric dating had found that a typical deep-sea clam a third of an inch across was a century old. The hot-springs clam, *Calyptogena magnifica*, was nearly nine inches across but just ten years old. In other words, its growth rate was 300 times faster than that of its stunted cousin. Clearly life processes were not always slow in the deep sea.

Corliss and Edmond's first sight of this oasis, in 1977, had been like a mirage. Searching for a hot spring, they had stopped *Alvin* to pick up some volcanic rocks. As the pilot maneuvered the submersible's mechanical arm, Corliss and Edmond gazed with detached interest—after all, one was a geologist, the other a geochemist—at a couple of large purple sea anemones. Then, looking beyond the anemones, they saw that the water illuminated by *Alvin*'s spotlights was shimmering—"like the air above a hot pavement," Edmond later wrote in *Scientific American*. They drove up the slope.

> Here we came on a fabulous scene. The typical basaltic terrain at the ridge axis is bleak indeed. Monotonous fields of brown pillows are cut by faults and fissures. One must examine several square meters to find a single organism. Yet here was an oasis. Reefs of mussels and fields of giant clams were bathed in the shimmering water, along with crabs, anemones and large pink fish. The remaining five hours of "bottom time" passed in something close to a frenzy. . . . We worked until the "scientific power" ran out.

The sheer mass of life was stunning even to a geologist—stunning to anyone who had seen the ocean floor, in *Alvin* or in Heezen's picture book, stunning to anyone who had seen the desert but never an oasis. The question raised by an oasis is an obvious one: What is it that allows this miracle to exist here? In a desert on land the missing ingredient supplied by an oasis is water, but in the deep sea there is no shortage of that. Life is sparse in the deep sea because food is sparse, and food is sparse because sunlight is absent altogether. Without sunlight there can be no photosynthesis and no plants, which are—were then, in 1977—the base of every known food chain.

The ocean floor as it was known then was an entire ecosystem subsisting off the table scraps of another ecosystem—off the food falling from the sunlit ocean surface. Though a sudden fish fall might be enough to generate a passing cluster of scavengers, that was clearly not what was happening at the Galápagos hot springs. These communities were not communities of scavengers, for the most part; they were dominated by clams and mussels and tube worms that were rooted to the spot. What were all these animals finding to eat?

Even after the biologists' cruise in 1979 there would be debate on the subject. Perhaps, some researchers suggested, it was merely that currents flowing into the rising columns of hot water were concentrating the surface food fall above the springs. But the beginnings of the correct and more astonishing answer emerged as soon as John Edmond opened the first containers of water and animals that he and Corliss had collected with *Alvin*. The shipboard laboratory was immediately filled with the stench of rotten eggs: hydrogen sulfide. There are organisms that can live off hydrogen sulfide; they are called sulfur bacteria. They had been known to science for nearly a century. What was not known was that such bacteria could serve as the base of an ecosystem—the first ecosystem ever found on Earth that derived its energy, not from sunlight, not from photosynthesis, but from chemosynthesis.

The two processes are fundamentally analogous. In both cases organic carbon compounds—carbohydrates—are synthesized from inorganic carbon dioxide. Besides carbon dioxide, both processes require a source of energy to power the carbohydrate synthesis, a source of electrons to carry the energy, and an "electron acceptor" to catch the electrons when they have done their job. In photosynthesis the energy comes from sunlight: A photon of light raises an electron in chlorophyll—the green plant's sunlight-catching molecule—to a higher energy. As the electron cascades to a lower energy like water through a turbine, it drives enzymatic reactions that make ATP, the universal energy currency—the electricity, if you will—of all living cells. At the bottom of the cascade the electrons are accepted by another molecule, NADP, and eventually they help reduce, or deoxidize, carbon dioxide to carbohydrate. The cell restocks itself with these valuable electrons by splitting water molecules apart. In the process it spits oxygen out into the ocean or the atmosphere for animals like us to breathe.

Most chemosynthetic bacteria need the oxygen produced by plants too, but other than that they are completely independent of the sunlit world. They use oxygen rather than NADP as their electron acceptor. Their source of both electrons and energy is a chemical compound that has both to spare—a compound that is already reduced, like hydrogen sulfide. Sulfur-eating bacteria oxidize hydrogen sulfide, stealing its electrons and latent energy and using them to reduce carbon dioxide. They can only live in places where they have access to both hydrogen sulfide and oxygen. The combination is not so easy to come by. One place the sulfur eaters can have the best of both worlds is in coastal mud flats. There they get hydrogen sulfide from other mud-dwelling bacteria, which make it from sulfate in seawater, and oxygen from the seawater itself. Until 1977 biologists regarded sulfur eaters as mud-flat curiosities; certainly not as world builders.

Yet when the divers in *Alvin* found that the shimmering water at the Galápagos hot springs was milky with these bugs—a milliliter of it, a cube one twenty-fifth of an inch on a side, contained a million cells or more—most researchers quickly concluded that chemosynthetic bacteria must be the food that nourished the new ecosystem. Certainly, as Edmond had discovered, there was plenty of hydrogen sulfide in the vent water: as seawater percolated down through cracks in the splitting crust, the sulfate in it was reduced to hydrogen sulfide by chemical reactions with the hot volcanic rock. But this interpretation of the hot-springs communities immediately ran into a problem. For the most part the densely clumped organisms, the giant clams, mussels, and tube worms, lived at some distance from the springs themselves, and thus from the warm, bacteria-laden water. It was hard to see how those sedentary organisms could be sifting enough microbes from the water to be growing to their prodigious size.

The conundrum grew far worse when biologists got their first close look at the giant tube worms, *Riftia pachyptila*. The tube worms grew in the densest clumps of all—indeed, they formed the densest concentration of biomass ever found anywhere on the planet—but they turned out to have a startling deficiency: they had neither a mouth nor a gut. In other words, they had no way of eating bacteria even if there were enough around to feed them. The tube worms were so different from known animals they constituted an entirely new kind of organism, not merely a new species or genus or family but a new subphylum, and some would say even a new phylum. Human beings belong to the subphylum of vertebrates, which includes everything from goldfish to tyrannosaurs, and to the phylum of chordates, which includes things like sea squirts as well. Thus to say that *Riftia* belongs to a new phylum is to say that it has less in common with everything else that lives in the sea than fish or sea squirts do with us.

It was the tube worms above all that made deep-sea biologists realize that they were onto something even more surprising at the Galápagos than they had first thought. Astonishing as the notion was of a whole community of animals feeding on chemosynthetic bacteria, of a deep-sea ecosystem supported not by trickle-down sunlight but by trickle-up geochemistry, it was not quite astonishing enough.

———————

Colleen Cavanaugh was just a first-year graduate student when she had her epiphany, the sort of flash of insight scientists are lucky to get once a career and most never get at all. She had come to Harvard from an undergraduate career at the University of Michigan, where she had dabbled in music but had ended up majoring in ecology under the tutelage of a man who specialized in coral reefs. From him she had learned that coral reefs are founded on a symbiosis: the individual animals, called polyps, harbor microscopic plants called zooxanthellae in their tissues. The animals provide nutrients to the plants, which in turn harness the energy of sunlight to make organic matter that feeds the animals. Later, as a research assistant at Woods Hole, Cavanaugh had worked with microbial ecologists who were studying sulfide-making and sulfide-eating bacteria in coastal sediments. When she entered Harvard in the fall of 1979, just a few months after the first biological cruise to the Galápagos vents, she had, as she puts it, "sulfur bacteria on the brain." It turned out to be the right place and time for someone with that affliction.

"The vents had been discovered while I was in Woods Hole, so I knew about them," Cavanaugh recalls. "In my second term as a graduate student, we had a course called 'Nature and Regulation of Marine Ecosystems.' There was a series of talks on the vents, and the final talk was one on tube worms given by Meredith Jones, who at that time was the curator of worms at the Smithsonian. Since the animals have no mouth and no gut, the question was, how are they feeding at the vents? Anyway Jones gave this lecture, and it was kind of a micron by micron dissection of a six-foot-long tube worm. I mean it was unbelievable—slide after slide of anatomy, both at the cellular level and at the gross level. At one point he showed a cut through the worm that showed this brown spongy tissue that filled the body cavity of the animal—the trophosome. No one knew its function. But Jones mentioned that he had found all these crystals in it, in a number of the animals he dissected. He had the crystals analyzed, and it turned out they were elemental sulfur.

"It was at that point that I jumped up and said 'Well, it's perfectly clear! They must have sulfur-oxidizing bacteria *inside* their bodies, just like corals have zooxanthellae that are photosynthetic. The bacteria are oxidizing hydrogen sulfide'—elemental sulfur is an intermediate in the oxidation process—'to produce energy, fix carbon dioxide, and feed the animal internally. And all the animal has to do is transport the sulfide, the oxygen, and the CO_2.' It just made all the sense in the world! But Jones said 'Sit down, kid. We think it's a detoxifying organ.' Or words to that effect. Hydrogen sulfide, you see, is a potent toxin—it messes up humans as well as invertebrates, even bacteria. And I said 'That's fine! The bacteria are taking the toxin and oxidizing it to non-toxic forms like elemental sulfur and all the way to sulfate!' It truly seemed perfectly clear to me.

"Afterwards I talked to Jones and convinced him to send me some trophosome tissue, and I would prove there were bacteria there. Now, you have to understand, I was a first-year graduate student, I'd never had biochemistry, I'd never had a microbiology class. So the first thing I did was look at the tissue with a scanning electron microscope. I saw what looked like grapes. The whole tissue just looked like these round things. That discouraged me—to me bacteria looked like *E. coli*, like rods. I went and talked to Holger Jannasch [a deep-sea microbiologist at Woods Hole who had been among the first to suggest that chemosynthesis was happening at the vents] and he said 'Well, remember, bacteria when they're inside other cells, can lose their own cell walls and become quite odd-shaped.' So that was one hint."

Cavanaugh went on to find other hints. She showed that DNA was spread throughout the "grapes," which is what you would expect of bacteria but not of tube-worm cells, which like all animal cells confine their genetic DNA in a small central nucleus. She also determined that the trophosome tissue was rich in lipopolysaccharides, a type of molecule found in certain bacteria but not in animal cells. In the end she proved her case to Jones's satisfaction and also to Jannasch's; the three of them and two other researchers wrote a paper together describing their discovery of symbiotic bacteria. The paper appeared in *Science* in 1981, along with five other papers on *Riftia*. It was a lot of pages for the most prestigious American science journal to devote to a single deep-sea worm. But then, *Riftia* was a very strange worm. It was the first animal ever discovered that was nourished by endosymbiotic, chemosynthetic bacteria.

It was far from the last. The giant clam, *Calyptogena magnifica*, soon turned out to have gills stuffed with sulfur bacteria; so did the mussel *Bathymodiolus thermophilus*. Unlike *Riftia*, these mollusks are not totally dependent on their bacterial friends for food; to a certain extent they can filter organic matter from the passing water. But without their endosymbionts they would probably not be able to survive at the hot springs, let alone grow in such abundance and to such size. These three types of organisms—tube worms, clams, and mussels—make up the bulk of the living matter found at the Galápagos and at many other hot springs that were discovered later. The hot-spring ecosystem is thus founded on chemosynthesis, first of all, but also and just as important on endosymbiosis.

Exotic as this new ecosystem was, it soon began to seem less so. Once *Riftia* had opened their eyes, Cavanaugh and other researchers began to see chemosynthetic symbionts in far more familiar places. "*Riftia* was the first," Cavanaugh says, "but then the question was, Why would this only occur at hydrothermal vents? There's sulfide in marshes and coastal sediments. Then a paper came out describing a gutless bivalve, *Solemya reidi*, from the West Coast, that lived in sulfide-rich sediments associated with pulp mills. And I said, 'Wow, It's perfectly clear! It's gutless because it has symbionts.' There is another species of *Solemya* here on the East Coast, *Solemya velum*, and so I went and looked at that one. This is an animal that had been described since the 1800s, and its whole natural history was curious. First of all, it has a very small gut. It's supposed to be a deposit feeder—it's supposed to be a protobranch, which are the most primitive type of bivalves—but it was unusual in that its feeding palps, which protobranchs send out to bring in the sediment, had degenerated so much they didn't even reach outside the shell! It lived in a Y-shaped burrow, at the apex of the Y, and it pumped water from above, and had an in-current siphon from below. That too is unusual for bivalves. So nothing fit. But if you postulate that these clams have sulfur bacteria in their tissue, it's perfect. They're getting the oxygen in the water in the upper part of their burrow, through the arms of the Y, and they're pumping the sulfide from below. It was perfect! I figured I didn't even need to do the experiments, but of course I went and looked. The gill tissue was packed with rod-shaped bacteria."

In the years since, Cavanaugh and others have found more than one hundred different species of marine animal, in five different phyla, that depend to some degree on a symbiosis with sulfide-oxidizing bacteria. What all these organisms have in common is that they live at the interface between oxygen-rich seawater and sulfide-rich water seeping up from the sea-floor. The heat at hot springs is secondary; the important thing is that they are sulfide springs. Running seawater over hot volcanic rock is just a particularly efficient way of making a lot of sulfide.

In 1984 biologists discovered there are other ways. Diving

in *Alvin* to the base of the Florida Escarpment—a mile-high cliff that plunges from the continental shelf onto the two-mile-deep abyssal plain of the Gulf of Mexico—Barbara Hecker of Lamont-Doherty and Charles Paull of Scripps discovered dense communities of mussels and tube worms clustered around patches of black sediment. When the researchers brought samples of the sediment onboard the ship, they could smell it: hydrogen sulfide. Seawater apparently percolates deep into the fractured face of the limestone cliff, and there bacteria make sulfide from the sulfate in the water. The heavy, sulfide-rich water is then squeezed out at the base of the cliff, where it nourishes giant tube worms—indirectly, by feeding their symbiotic bacteria. The temperature at the base of the Florida Escarpment is just what it is elsewhere on the floor of the Gulf of Mexico, around 40 degrees Fahrenheit. In other words, the animals at the Florida Escarpment are supported not by a hot spring but by a cold seep.

If the heat is not the essential element in hot springs, nei-ther, in a sense, is the sulfide. The mussels at the Florida Escarpment have symbiotic bacteria in their gills, but those bacteria are not sulfur eaters. They are methane eaters: they use the energy they get from oxidizing methane to make car-bohydrates. The methane is also made in the limestone cliff, by bacteria digesting the remains of ancient organisms. Else-where in the world—just like hot springs, cold seeps have been discovered in many places—the methane may have a different source. In deep-sea trenches off the coasts of Japan and Oregon, where one tectonic plate dives under another, the collision squeezes methane out of the seafloor mud. In the northern Gulf of Mexico, off Louisiana, the methane and sul-fide escaping from the seafloor come from the same oil depos-its that human beings have exploited so aggressively. Like us, the mussels and tube worms off Louisiana—or rather their bacteria—are burning fossil fuels.

The essential element of both hot-spring and cold-seep

communities, then, is a copious supply of a reduced chemical, either sulfide or methane, that can serve as a fuel for the animals' symbiotic carbohydrate factories. The other essential element is oxygen. Just as a fire will go out when the oxygen is gone, so too will the bacteria stop oxidizing sulfide or methane. They can only survive in the narrow boundary zone, where oxygen-rich seawater mingles with the sulfide- or methane-rich water emanating from the seafloor. It is a constantly shifting front, and the organisms have to find some way to bridge it.

This creates a problem for the bacteria, especially for the sulfur bacteria, because sulfide oxidizes spontaneously as soon as it comes into contact with oxygen. It is the problem we human beings would face if all the oil we pumped so furiously and all the coal we mined so industriously were to burst into flame the moment we succeeded in bringing it to daylight. Sulfur-eating bacteria want the burning to take place inside them, so they have to find a way to get to both their fuel and their oxygen before the two meet on their own. To Cavanaugh, who is after all a microbiologist, the glorious jungles of animal life we see today at seafloor vents are but side effects of the way her inconspicuous bugs have contrived to solve this fundamental problem of theirs.

Sulfur bacteria that live free in seawater or sediment have all sorts of tricks for getting the two elements they need. Some of them migrate back and forth with the sulfide-oxygen front, trying to straddle the fence at all times. Others let the front surge past them but keep a supply of pure sulfur as a "lunchbox" (Cavanaugh's term) to tide them over when times are lean: when they cannot get any sulfide they burn the sulfur, and when they cannot get any oxygen they use the sulfur to oxidize the sulfide. Still other bacteria hold onto the front by forming a thin white veil over a dead animal, trapping the sulfide emanating from the corpse below while taking in oxygen from the water above.

But the most ingenious strategy of all is the one adopted by bacteria that are too small or too slow for any of these tricks: "Jump on anything bigger than you," as Cavanaugh puts it. Not quite any animal will do as a bacteria vehicle, of course. It has to be able to migrate with the sulfide-oxygen front, the way some worms that have endosymbiotic bacteria have been found to do. Or it has to be able physically to span the front, the way a *Solemya* clam does with its Y-shaped burrows or the way the vent clam *Calyptogena* does: it lodges itself in a crack near a hot spring, collecting sulfide-rich water from the volcanic rock with its large, probing foot, while siphoning in oxygen-rich water from the sea. Or, finally, the animal has to be able to span the sulfide-oxygen front in time, the way the tube worm *Riftia* does. It takes in water that is now rich in sulfide, now rich in oxygen and channels both to the busy symbionts in its trophosome.

In the various types of symbiosis that exist today Cavanaugh sees evolutionary stages, all beginning with a time long ago when sulfur bacteria first decided to jump deep-sea animals. Nematode worms have been found that graze on the surface of their own body, which is covered with chemosynthetic bacteria; that is the first stage of the bacterial invasion. Other worms keep their bacteria just under their skin but outside their cells; that is the next stage. In clams such as *Calyptogena* the bacteria have entered the cells, but only in tissue such as the gills that carry out additional functions besides chemosynthesis.

In *Riftia* the evolution is complete, and the bacteria have won: the trophosome is first and foremost a home for them, and the worm has no other source of food but what they make for it. In return the worm has adapted its very chemistry to meet their needs. It has evolved a form of hemoglobin, similar to the molecule that transports oxygen in our own blood, but one that can transport both oxygen and hydrogen sulfide, all the while keeping the two at arms length until they can be delivered to the bacteria. It is this hemoglobin that makes the blood of *Riftia* red like ours, and that gives it the blood-red plumes, so beautiful and so extraordinary in a deep-sea animal.

———

Cavanaugh does not know, nor does anyone else, when and exactly how this evolution might have begun. For that matter, it is not at all certain how the deep sea came to be inhabited at all, at vents or anywhere else. Are the abyssal animals migrants from shallow water? Or vice versa? Or both? In the case of the vent communities, though, there are reasons to believe they have been around in much their present form for some time. One reason is that around a decade ago, Rachel Haymon found fossils of 95-million-year-old tube worms in a copper mine in northern Oman.

If seafloor hot springs were nothing more than oases for exotic animals, they might never have been found; it was, after all, geologists and geochemists who went looking for them. The hot springs are also furious chemical reactors, in which seawater reacts with rock under great heat and pressure. The conversion of sulfate to sulfide is the reaction that is important to the organisms at the vents, but it is just one of many that take place as the water percolates down through a mile or more of fractured crust, is heated there by molten rock, and then bubbles back toward the surface. The water becomes so hot and so acid that it dissolves metals—copper, iron, and zinc—out of the rock it passes over. At the first hot springs off the Galápagos, researchers did not get a chance to see the effects of all these reactions, because the hot springs there were not so hot—only 70 degrees Fahrenheit or so. That is warm indeed for the deep sea, but not nearly so warm as water that has just been heated by magma. Apparently the water bubbling up off the Galápagos gets mixed with cold seawater and cooled before it reaches the seafloor.

But in 1979 geochemists like Haymon, who was then a graduate student, learned what happens when vent water does not get cooled on the way up:It comes out of the seafloor at 350 degrees Celsius, which is 662 degrees Fahrenheit, or more. The researchers discovered this on a cruise led by Fred Spiess of Scripps and Ken Macdonald to the East Pacific Rise at a latitude of 21 degrees north, just outside the mouth of the Gulf of California and 1,800 miles north of the Galápagos. One day the divers in *Alvin* happened on a slender, six-foot-tall column of rock that was spewing what looked like black smoke; the pilot compared it to the smokestack of a locomotive. He inserted *Alvin*'s thermometer into the smoke to measure its temperature. The plastic tip of the probe melted.

The smoke turned out to be sulfides of iron and other metals that were precipitating out of the hot solution as they erupted into the icy seawater. The metal sulfides were precipitating so fast they had formed a solid chimney of rock that was growing upward like a stalagmite, only much faster. But most of the particles escaped the top of the chimney in billowing clouds. They fell like snow on the surrounding rift valley, covering it with a carpet of frozen smoke. The carpet was incredibly metal-rich: it was 31 percent zinc, 14 percent iron, and 1 percent copper.

At first it was thought that black smokers would only grow in the hot and fast-spreading Pacific, but in 1985 they were found on the Mid-Atlantic Ridge as well. They have now been discovered all over the world. (The largest yet found is on the Juan de Fuca Ridge off the coast of Washington: it is called Godzilla, and it is 16 stories tall.) The smokers' effect on the chemistry of the ocean is profound. It has been estimated that an amount of water equivalent to the entire world ocean flows through hot springs on midocean ridges every ten million years. That works out to one Amazon River's worth of water in a single year. A chemist's calculations of where the various elements in the ocean come from and where they go will not work out well if she ignores the Amazon.

Ocean chemists knew something was missing from their

calculations before black smokers were discovered, but they were not sure what. The effects of the smokers had been unwittingly observed on land, however, and had been exploited for centuries, not by chemists but by miners. Over geologic time black smokers, steadily filling rift valleys with their metal-liferous smoke, have created some of the planet's richest ores.

The classic example is Cyprus. The word "copper" comes from *Kypros*, which is the Greek name for an island that has been famous since antiquity for its copper mines. Thanks to the theory of plate tectonics and thanks to the discovery of seafloor hot springs, we now know why. The copper mines on Cyprus tap an 800-square-mile block of ore called the Troodos Massif. It is what is known as an ophiolite: a vertical slice of midocean ridge that was beached on land when two plates collided and closed an ocean. The ocean in this case was the Tethys Sea, an east-west gash that split apart the supercontinent Pangaea 200 million years ago; the Mediterranean is a tiny remnant of it. The Tethys survived until a younger ocean, the Atlantic, sundered the continents in a north-south direction and began shoving Africa and Eurasia back together. One trace of that collision is a band of ophiolites that arcs from Cyprus and Turkey through Iran and across the Persian Gulf into Oman. In one of those ophiolites, in northern Oman, there is a copper mine called the Bayda mine. That is where Haymon found her fossilized tube worms.

The metal sulfides that Haymon was examining had been deposited by a Tethyan black smoker 95 million years ago, and in the process they had buried the worms that lived around the smoker. The fossil tubes were unmistakable. They were similar to sulfide-encrusted *Riftia* tubes that Haymon had already found at 21 degrees north in the East Pacific—or, for that matter, to the freshly killed fossils-in-the-making she would later see at 9 degrees north. To be sure, the Bayda tubes were much smaller than *Riftia*—less than half a centimeter in diameter—and in that respect they seemed more closely related to a species of tube worm that still lives on the Juan de Fuca Ridge. In any case, the fossils demonstrated that tube worms of some description had been living at seafloor hot springs since the Late Cretaceous Period, when dinosaurs still stomped the Earth. In their splendid seafloor isolation the tube worms had managed to survive the mass extinction that had wiped out the dinosaurs and other surface dwellers at the end of the Cretaceous, 65 million years ago.

There is another reason to believe that hot springs have been inhabited for a long time, perhaps even since long before the Cretaceous. The reason is simply that the animals are so strange. Nearly every one of them was new to science when it was found at a vent; by and large things that live at vents cannot live elsewhere in the ocean, and vice versa. The vent animals are not just different species from other deep-sea animals; in many cases biologists have had to create new families, orders, or even, as in the case of the tube worms, new phyla to accommodate their strangeness. To have become that different from other deep-sea animals in a short time, they would have had to evolve very fast. It seems more plausible to believe instead they have evolved at no more than an ordinary rate, but have been isolated in their own world for a very long time.

Indeed, some researchers have suggested the vent lineage is of extraordinary antiquity: that life began at a seafloor hot spring. One of the most vigorous advocates of this theory has been John Corliss, who was also one of the first humans to see a hot spring. Like so much that has been written about the origin of life, Corliss's theory is really only a speculation, and it is not likely to be proved or disproved anytime soon. But it is a fascinating speculation, one that has at least two things going for it—two advantages over the old Miller-Urey theory of a primordial soup.

The first advantage concerns the conditions required for making the organic precursors of life. The reducing atmosphere that was invoked by the primordial soup theory, to keep simple molecules from being oxidized before they could be combined into more complex organic ones, is no longer believed to have existed on the early Earth—but reducing conditions certainly exist in the interior of a seafloor vent. As Corliss sees it, molecules such as amino acids could have been synthesized at great heat and depth in a vent, and then rushed to the surface and quenched in cold seawater before the heat had time to rip them apart again. The molecules would have accumulated on rocks at the mouth of the vent. There they could have evolved into the first polymers and, eventually, into the first fragile cells.

The inevitable fragility of these early attempts at life is where the second advantage of the vent scenario comes in. While the surface of the early Earth was being peppered with meteorites large and small, the bottom of the ocean, Corliss thinks, may have offered the only safe refuge on the planet. At a hot spring, the first organisms could have found nourishment as they survived the rain of rock, just as their descendants would later survive the single asteroid impact that is thought to have wiped out the dinosaurs. From their seafloor sanctuaries the primordial organisms could then have colonized the rest of the deep and the rest of the planet.

If hot springs have indeed been a sanctuary on a hellish Earth, they have certainly been a harsh one. It is not just the capricious gusts of 650-degree water that scald animals even when they have kept a careful distance from the vent. Nor is it the omnipresent sulfide that is at once food and lethal toxin. It is not even the black smoker chimneys that are always growing to the limits of structural soundness, then collapsing on the animals below. No, the harshest thing about hydrothermal vents as a home is their essential temporariness. The heat from that particular body of magma dissipates, the cracks in the seafloor heal, and the vent winks out, just like that. No one knows for sure how long a given vent lasts, but it is probably on the order of decades. At the Galápagos in 1977, at the same time that geologists were discovering the first vents teeming with life, they were also discovering the first dead ones— patches of seafloor where the shimmering heat had stopped, the tube worms had decayed or been buried, and all that remained was a field of once magnificent clam shells, slowly dissolving in the acidic water. Those animals' home was gone, and they had no prospect of finding a new one. Their only hope, in an evolutionary sense, was that during the short period of bounty they had managed to spread their seed: to spawn offspring that would land in a part of the seafloor where random geologic forces had chanced to create a future.

All deep-sea animals live in a patchy environment, but only the vent creatures face such a crisp line between the conditions of survival and those of death, and such an urgent need to grow and reproduce fast—only the vent creatures, that is, and one other group. This other group was known before the vents were discovered. In 1972, on the same series of *Alvin* dives on which Fred Grassle was finding that his sterilized trays of deep-sea sediment were recolonized only very slowly by worms and clams, a biologist named Ruth Turner from Harvard was conducting a different experiment. Turner specializes in boring mollusks—mollusks that tunnel into wood or concrete and thereby do extensive damage to wharves and ships. (Shipworms are one family of boring mollusk.) Instead of sediment trays, she used *Alvin* to stand foot-long pine boards upright in the deep-sea mud 110 miles south of Woods Hole. When she retrieved the boards three and a half months later, they were riddled with holes like a honeycomb. More than riddled: as *Alvin*'s mechanical arm gripped them, the boards started to crumble. Tens of thousands of minute bivalve mollusks, of the family Xylophagainae, had gnawed their way into the boards from both sides. Rhythmically opening and closing the two valves of their shell, they had rasped the wood fibers to dust with the serrated edges of the valves. The wood was not just a home for these queer mollusks but also food—food that nourished their long wormlike bodies, only the front end of which was enclosed in the shell. (The rear end reached out the mouth of the tunnel into the water, where it could collect oxygen.) In fact, wood grown on land was what they were specifically adapted to eat, even though they lived at a depth of 6,000 feet on the seafloor.

Theirs was the chanciest of existences. Although there might well be a seasonal cycle to their lives—spring is when large amounts of deadwood gets washed off the land and out to sea—there was no predicting when or where a particular log or branch might become waterlogged, sink, and enter the domain of the Xylophagainae. Certainly there was no wood in sight when Turner deployed her boards. And yet within just a few months the mollusks had managed to find Turner's wood, settle on it, and eat their way to adulthood. Incapable of swimming, they were doomed to die once their serendipitous food source was exhausted; having once gripped the wood with their strong sucker-foot, they would never leave. But their progeny would. When Turner retrieved her crumbling boards, the mollusks' gonads were nearly ripe. They were on the verge of spawning.

Turner has yet to figure out exactly how her wood borers go about spreading from one ephemeral island of food to another, but it is clear that the spreading is done by floating larvae. And so it must be with most of the creatures at hot springs—the sedentary adults must count on floating larvae or juveniles to find and colonize a new hot spring when the old one dies. Biologists are just beginning to study how this process works; the microscopic larvae of deep-sea invertebrates are hard to find, and when they are found they are hard to tell apart. But it is already possible to say something about how successful the process has been over long, geologic periods of time. One has only to look at the present distribution of the hydrothermal vent animals.

In a sense they live in a one-dimensional world. Their sole avenue of dispersal, or at least their principal one, is along the crest of the mid-ocean ridge. Larvae may well drift off the ridge, but they will almost certainly not survive. The narrowness of the vent creatures' world, however, has a flip-side advantage: it is their world and only theirs. Few animals from the rest of the deep, not specially adapted to the conditions at the hot springs, can brave the heat and plumes of toxic hydrogen sulfide to compete with the vent creatures. Moreover, although an individual hot spring may fizzle out at any time, at great loss of life to its inhabitants, their descendants can more often than not find another active vent within a few miles of the first. Thus over time, by hurling huge numbers of larvae into the water and suffering an unending Okinawa of losses, the vent animals have been able to conquer new territory. Indeed, over millions of years they have apparently island-hopped their way to the ends of the Earth.

From the Galápagos the oceanographers followed the hot-spring trail west and north to the East Pacific Rise. In the late 1970s and 1980s major vent fields were found along the rise at 9 degrees north, 13 degrees north, 21 degrees north (where black smokers were discovered), and even in the Gulf of California itself, just south of where the rise terminates in the San Andreas Fault. After terrorizing much of California the San Andreas reemerges from the continent near Mendocino and rejoins a spreading center 275 miles offshore, and there the hot-springs trail picks up again. Vent communities were discovered along this northern spreading center, the Juan de Fuca Ridge, in 1983. The Juan de Fuca was once simply the northern end of the East Pacific Rise, and vent animals could spread all along the rise off of California. But around 30 million years ago the continuity of the rise was broken by the westward moving North American continent, which was then and is still overriding the Pacific ocean floor, mountain ranges and all, and dunking it into the mantle. The gentle spread of the rise was transformed into the destructive shear of the San Andreas. And the hot-springs animals were cut off from their southern kin, the closest of which are now nearly 2,000 miles away.

The separation shows in the species composition of the two regions. The Juan de Fuca vents are dominated by tube worms, but instead of *Riftia* they belong to a much smaller and less showy genus called *Ridgeia*. There are *Calyptogena* clams up north, but they are far from *magnifica*; they are a much smaller species called *pacifica*. Verena Tunnicliffe of the University of Victoria has analyzed the Juan de Fuca fauna comprehensively and has found that more than 80 percent of the species that occur there occur only there and not on the East Pacific Rise. On the other hand, the two regions share many genera and higher taxonomic groups. They have been evolving separately for the past 30 million years, but apparently they have been evolving slowly: if they had been separated for longer or if their differences had been evolving more rapidly, the different species might have evolved into different genera or even families.

The Mid-Atlantic Ridge and the East Pacific Rise are still connected today but in a roundabout way. There is no direct connection around Cape Horn. To get from one region to the other, the vent organisms would have to go around Africa, through the Indian Ocean, south of Australia, and across the Pacific (or the other way around). Not surprisingly, the mid-Atlantic fauna is more different from that of the East Pacific Rise than the Juan de Fuca fauna is. No tube worms have been found at all on the Atlantic ridge, and no *Calyptogena* clams, large or small. The most conspicuous animal at the Atlantic hot springs are a strange species of shrimp with rudimentary eyes on their backs. The shrimp buzz in huge swarms around the vents; in the otherwise total darkness of the abyss, they may use their eyes to home in on a faint and still-mysterious glow that emanates from the hot springs. Shrimp of the same family have been found in the Pacific, indicating that there has not been a total lack of communication between the two regions. The shrimp and a few other organisms may have managed to take a shortcut between Pacific and Atlantic, through the open passage that separated North and South America until the Isthmus of Panama rose out of the sea around three million years ago. But this is sheer speculation.

The route the vent organisms took to reach the Mariana Trough in the western Pacific is even more of a mystery. The hot springs there are not connected at all to the midocean ridge system: they lie on what is called a back-arc spreading center, a geologic feature that occurs not where two plates separate but where they collide. On the other side—the east side—of the Mariana Islands from the trough, the Mariana Trench plunges to nearly 37,000 feet, as the Pacific plate dives under the Philippine plate. Heat rising off this collision zone builds the arc of volcanic islands, but it also cracks the crust in back of the arc, creating a small spreading center—small and ephemeral. Back-arc ridges come and go along the western edge of the Pacific plate, like brief sideshows of the larger tectonic drama. The one that formed the Mariana Trough is around five million years old. It is around 5,000 miles from an active midocean ridge.

Yet the animals that live at the hot springs of the Mariana Trough have clearly been in touch with the ridge system, or at least their ancestors have. When Bob Hessler and his colleagues went looking for and found the Mariana hot springs in *Alvin*, their first look revealed a strikingly different community from the ones on the midocean ridge. Instead of tube worms, clams, or shrimp, the vents were choked with large snails whose tightly whorled shells made them look hairy. The snails belong to an entirely new family: They are the first snails known to be nourished by symbiotic bacteria. But as Hessler and a squad of taxonomists analyzed the Mariana fauna, this type of uniqueness turned out not to be the rule. Of the 27 species that were positively identified, 13 belonged to genera known from midocean ridge vents in the eastern Pacific or

Atlantic, and another three to known families. So while the hairy snail and a few other species may be western Pacific natives that adapted to hot-springs life on their own, most of the animals at the Mariana vents trace their ancestry to the midocean ridge. That raises the question of how they got across five thousand miles of hostile open water.

No one knows the answer to that question. One possibility, according to Hessler and geologist Peter Lonsdale of Scripps, is that the vent organisms made their way to the western Pacific tens of millions of years ago by a route that no longer exists. Until 55 million years ago the Indian Ocean link in the midocean ridge passed north of New Guinea, not far from the Marianas (it now runs south of Australia). And until 43 million years ago there was an active ridge running right across the tropical Pacific, from the East Pacific Rise to Asia. Vent animals might have spread along either of these ridges to the western Pacific and survived at other back-arc basins until 5 million years ago, when the spreading center in the Mariana Trough opened for business. Or they might have come to the region more recently, by using spreading centers in other back-arc basins to the south and east of the Marianas as stepping stones. Those spreading centers are still active today, and vent communities have been found on them too. From the southernmost of them, the Havre spreading center south of Fiji, it is a mere 2,000 miles to the Southeast Indian Ridge, the nearest link in the midocean ridge chain. New Zealand lies directly in the way. Could vent animals have spread across that gap and around New Zealand? Could they still be doing so today? Again, no one knows.

Some things are known, however, about the spread of the vent animals today. For instance, in some places it appears to be very fast. Although the larvae themselves are difficult to track, Richard Lutz and his colleague at Rutgers, geneticist Robert Vrijenhoek, have measured the dispersal of the animals in a different way: by tracking the flow of genes between vents.

In any organism certain genes and the proteins they produce tend to come in various versions, called polymorphisms; the blood groups are a familiar example in humans. When different populations of the same species are separated, so that they are no longer routinely interbreeding, they evolve different constellations of polymorphisms. The degree of difference reflects the "genetic distance" between the populations and the amount of gene flow that is occurring between them. Since genes are carried by larvae, the rate of gene flow between two vent sites tells you something about the rate of larval flow.

If the same species is present at both sites, that alone indicates that some gene flow is occurring, or at least that it has stopped recently in a geologic sense; otherwise the two populations would already have evolved into separate species. But between the Galápagos and the East Pacific Rise and among the different vent fields on the rise, Lutz and Vrijenhoek have found truly astonishing amounts of gene flow. Not only are the same species of tube worms, clams, and mussels present at great intervals along the ridge crest, but in some cases organisms hundreds or even thousands of miles apart turn out to be all but genetically identical. That means one population is descended from the other in the very recent past, perhaps only decades ago.

Such rapid dispersal suggests there is a kind of highway in the deep—a larval highway. The vents may build part of the highway themselves. The hot water spewing from a vent rises off the seafloor like smoke from a smokestack, in a narrow, roiling plume, and as it rises it sweeps up and entrains the floating larvae of the vent organisms. Lauren Mullineaux and Peter Wiebe of Woods Hole have proved this is true by pulling larvae from a plume with fine plankton nets towed above a field of vents on the Juan de Fuca Ridge. That single vent field, they estimate, dispatches as many as 700 million larvae a day. Some of the larvae may be bound for a nearby vent or may even fall out of the plume before they leave home. Others are surely headed for more distant places. Toxic metals in the plume, which they can withstand, may offer them some protection from outside predators. The larvae may also be able to slow down their metabolism, the better to survive a long journey without food through a cold and inhospitable abyss.

Yet the vent larvae themselves are not in control of their fate. They do not swim, they drift; and just as they must depend on random twitches of the seafloor to create for them a home, so must they rely on deep-ocean currents to lead them to it. When the hot plume rising off a vent reaches an altitude of 700 to 1,000 feet off the seafloor, it has cooled enough for its density to equal that of the surrounding seawater. Then it begins to spread out horizontally, and the vent larvae spread with it. From then on they are at the mercy of the currents. If oceanographers understood how water flows in the abyss better than they do, if they had maps that showed in some detail the pattern of the currents, then much of the mystery about how animals spread from vent to vent and around the globe would be removed. But physical oceanographers are only just beginning to draw such maps; and for the moment, the biologists are still at the stage of describing how well the system works. All they can say is that somehow, when opportunity knocks on the mid-ocean ridge, the vent animals are there to seize it.

Since witnessing the birth of a hydrothermal vent field in April 1991, at 9 degrees north on the East Pacific Rise, Rachel Haymon and her colleagues have been back several times. Richard Lutz has tracked the biological side—documenting how a hot spring evolves, the way a forest has long been known to do, with one species of tree succeeding another. In March 1992, at a vent called the Hole to Hell, Lutz saw clumps of foot-long *Tevnia*—accordionlike tube worms—where eleven months earlier there had been only fresh lava. By his next visit, in

December 1993, the *Tevnia* were being crowded out by *Riftia*, some as much as five feet tall, that had arrived some time during the interim. Five feet in no more than 21 months would be a healthy growth rate on land; in the deep sea it was a record. All over the vent field at 9 degrees north the *Riftia* were spawning vigorously. Nevertheless, by late 1995 mussels were starting to encroach on the tube worms' turf, and Lutz was waiting for the giant clams to arrive.

There is much still to be learned at 9 degrees north about how hot springs evolve. But the most fundamental lesson was evident the first time the researchers returned, in March 1992, to the Hole to Hell and the Tube-Worm Barbecue and the Valley of the Shadow of Death, where Haymon and two companions had once bumbled around in the inky smoke. The lesson was that change on a midocean ridge can come with shocking velocity.

"The Tube-worm Barbecue looked like a totally different place," Haymon recalled not long after the 1992 expedition. "All the iron sulfide in that gray sediment had oxidized to orange, so the site was just blanketed in orange. When we had seen oxidized sediments around hydrothermal areas before, we had tended to assume they were old sites—tens to hundreds of years old at least—but obviously it does not take that long at all. In 1991 this area also had warm water pouring out of a gaping, 10-foot-wide fissure, cascading up the caldera wall like a waterfall flowing backwards. In 1992 there was no venting at the Barbecue. There were still tube-worm casings littering the floor, but there was not a shred of fresh tissue left. There were a lot of fish, little white octopuses, crabs. Anything that could move, that could swim in or crawl in, was there, and was consuming what was left of the tube worms. It looked like a completely different place."

Rich Lutz: "And the Barbecue was not nearly as interesting as the sites that were newly formed hydrothermal vents. At the Hole to Hell, where the water was coming out of bare basalt in 1991, when we went back there was a 16-foot-high black smoker. That's a pretty impressive rock to see that wasn't there 11 months earlier. And there were tremendous quantities of *Tevnia*; there were thousands of them in any given vent area."

Dan Fornari: "It was astounding the change that we saw. All of the areas that had been white, white, white with bacterial mats, and 'snow blowers' everywhere—there was now very little bacterial mat. All the little tiny crabs we saw before—the area was now covered with crabs that were big and fat."

Haymon: "So we realized that the bacteria were being eaten. Everywhere in the caldera there were lots of animals, enormous numbers of crabs, and they were feeding on the bacteria. The Valley of the Shadow of Death went from having no animals in 1991 to having zillions of crabs. We changed the name to the Valley of the Crabs.

"I'm not a biologist, but it makes a nice story, and it may even be true: When you have an eruption, it's almost like springtime on the seafloor. The bacteria go nuts; instead of a greening of the sea floor, it's a whitening of the sea floor. That then provides a food source that causes a population burst in all the grazers. And in addition there are geochemical changes in the hydrothermal system that are like a trigger to the organisms that it's spawning time. When you boil seawater, you inject onto the seafloor a vapor that has an unusually low salinity and an unusually high gas content—hydrogen sulfide and carbon dioxide. So I'm just thinking that the animals can tell what's going on—that they pick up wafts of this vapor and say 'Oh, Oh. There's going to be an eruption, or there's been one. It's time to repopulate.'

"I don't know whether that's true. But definitely a lot of them got the message."

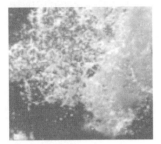

Blue Water

In 1969 Bill Hamner jumped off a ship in the middle of the Gulf of California, far from shore, into thousands of feet of blue water. He was still an ornithologist at the time, but he was getting fed up: he had developed an allergy to feathers. Walking into his aviary at the University of California at Davis was causing him to break out in hives. "I knew I wanted to change disciplines," Hamner recalls. "So I convinced these people who had a ship going to the Gulf of California to give me a berth. The ship was the *Proteus*, run by the Stanford oceanographic group, and I joined it that summer. One day while they were doing a lot of deep-water sampling, and the boat was sitting in flat-calm water, I jumped off the side— mostly just to cool off and because I was bored. But I put on a diving mask, and I saw that the ocean was just full of these gelatinous animals. So I convinced a couple of the guys to go on a scuba dive, just off the side of the ship.

"When you're diving in blue water, if you drift down away from the surface, you can't see the surface—you don't see anything but blue all around you. I was paying close attention to something, and I wasn't paying attention to my buddies. And all of a sudden I noticed that it was quite dark. I looked at my depth gauge, and it was well below 200 feet, and there was no bottom below me. I thought, 'Jesus Christ, this is a good way to die.'

"So I swam back up and collected my buddies. And while we were on the dive, the wind had come up, and the ship had drifted. When we came up to the surface it was two miles away. They had been keeping an eye on us, and came back to get us, but if the weather had been rough they wouldn't have found us. It scared everybody because it was so stupid. But we didn't know it was stupid."

After that first dive into blue water, Hamner resolved never to try it again until he had figured out how to do it safely. He knew, though, that he would have to try it again. The animals he had seen just by jumping off a ship had been too strange, too otherworldly even, and too beautiful; too luminous, diaphanous, and multifarious. Moreover, they did not have feathers—or even backbones for that matter. Medusae and ctenophores, tunicates and siphonophores, they all had one thing in common: they were made of jelly.

In 1890, just three sentences into his first report to the Berlin Academy on the results of his pioneering plankton expedition, Victor Hensen delivered himself of the following statement:

45. 70. 30. 45. 25. 20. 20. 35. 35. 65. 15. 45. 50. 20. 65.
40. 25. 30. 40. 25. 30. 30. 45. 45. 35. 30. 50. 25. 15.

To Hensen, a professor of physiology at the University of Kiel, those two lines of numbers said it all; they proved that his expedition had been a success. The numbers represented the volumes of plankton, in cubic centimeters, that his silk net had brought up at 29 stations between Bermuda and the Cape Verde Islands. They showed, Hensen claimed, that plankton was more or less evenly distributed in the open ocean. Contrary to what everyone had assumed before the expedition, there was a great deal more plankton in the colder waters of the North Atlantic than there was in the tropics, but otherwise the ocean was a pretty homogeneous place. With a few dips of the net you could learn something about the whole thing.

Provided, that is, you were willing to spend considerable time counting. Twenty-one years later, when Hensen finally published the full report of the expedition—*Life in the Ocean from Counts of Its Inhabitants*—there were tables and tables of raw numbers, and not just volume measurements. Hensen and his colleagues had spent years pipetting small samples from each haul of the net, and then counting, under a microscope, each of the thousands and thousands of plankters contained even in a small sample, and sorting them as much as possible into known species. "The undescribed species made for certain difficulties in the counting," Hensen lamented on the first page of his book. In Kiel at the end of the last century, there was a whole lot of counting going on—bean-counting, some would say, and some did, most notably the great zoologist Ernst Haeckel.

The origins of Hensen's interest in plankton are a bit obscure—his early career was devoted primarily to the study of human physiology—but it is clear enough that his interest was mostly practical: he wanted to know how much plankton was in the ocean in order to gauge how many fish it could support, the ultimate goal being to expand German fisheries. (Hensen had grown up in the Baltic coast city of Schleswig, not far from Kiel.) Haeckel, on the other hand, left no doubt about how he had become a zoologist. Near the outset of his extended critique of Hensen's expedition, he described his formative moment: it occurred in 1854, on a visit to the North Sea island of Helgoland with his mentor, Johannes Müller, inventor of the first plankton net. "Never will I forget," Haeckel wrote, "the astonishment with which I first beheld the swarms of glassy pelagic animals that Müller, by inverting his 'fine net,' emptied into a glass jar of seawater—that blooming confusion of dainty medusae and glistening ctenophores, of arrow-fast sagittas and snakelike tomopteris. . . ." Words like "astonishing," "fantastic," and "glistening" fell easily from Haeckel's pen, for all his learnedness; his favorite seems to have been *wunderbar*. And Hensen, he seemed to feel, was insufficiently alive to the wonders of the sea, because he had not studied them the way Haeckel had. He had counted them but not seen them.

By 1890, Haeckel had spent nearly four decades following Müller's method of "observation and reflection," collecting individual animals from a rowboat, ladling them carefully into glass jars, watching them and drawing them from life whenever possible, and above all describing, describing, describing, for there was so much to describe. The new species that annoyed Hensen were what Haeckel lived for. He had devoted a full decade to describing and drawing all the radiolarians that had been collected by the *Challenger*; his results ran to more than 1,800 pages of text and a separate volume of exquisite illustrations. Then he had devoted another few years to the siphonophores and sponges. He had just finished this great labor when Hensen set sail on his plankton expedition, without consulting Haeckel at all and with unheard-of support from the German government. Haeckel, who had spent his career in rowboats, stayed behind while Hensen steamed up and down the Atlantic for nearly four months in a generously equipped ship. One can imagine the envy Haeckel must have felt, the bitter envy, and also the anguish at being left on shore as the discipline he had given his life to sailed off on what seemed to him a completely wrongheaded course—as Hensen, in a sense, sailed off into the twentieth century without him.

No wonder Haeckel lashed out. "It seems to me," he wrote, "that Hensen has uncautiously drawn far-reaching and erroneous conclusions from highly insufficient experience. . . . I am convinced that his whole method of studying plankton is completely useless. . . ." Haeckel was an intense and passionate man, an artist as well as a scientist, and given to vigorous expression of his views. Hensen, on the other hand, appears from the distance of a century to have been a bit of a dweeb, albeit a steadfast and reliable one. Earnest, quiet, unapproachable until you somehow managed to become his friend, he was passionate about his measurements. One of his friends later insisted, as a sort of afterthought in a long memorial tribute, on Hensen's "warm feelings" for the beauty of the sea: had he not once remarked on the loveliness of the evening, that day he first tested his plankton-sampling methods?

Hensen's warmest feelings, though, were for the methods themselves. His savage response to Haeckel's attack on them makes that plain. In sixty pages of relentless sarcasm and boiling rage, Hensen tried essentially to excommunicate his antagonist, accusing him of fraud and deceit ("you can never trust Haeckel!"), of fantasy and thickheadedness ("absolutely every thing he says against my method is untenable"), of an obsession with large animals, and above all of the cardinal scientific sin: trying to impose his ill-founded theories on nature—and on other scientists—rather than grounding them in facts. "People tell me I will make too many enemies by fighting the Haeckel party," Hensen concluded. "So be it! . . . Only facts matter, facts that have been fixed as carefully as possible by weight, measure, and number. . . ."

Scientists these days do not throw firebombs with quite such abandon (more's the pity), preferring instead the silent shiv; but the Hensen-Haeckel debate is still smoldering in biological oceanography—this in spite of the fact that both men were both right and wrong, as G. Richard Harbison, a marine biologist at the Woods Hole Oceanographic, has pointed out. Hensen was right and Haeckel was wrong about the broad distribution of plankton: it is indeed most abundant at high latitudes. More generally, Hensen was surely right about the usefulness of quantitative methods. Although his own were primitive, their intellectual descendants have proved themselves again and again in the twentieth century. Without them, we could not understand as much as we do about how the chemistry of the ocean affects the life in it, how energy moves through the different levels of the food web, and how the ocean functions as an ecological machine. Most biological oceanographers today are Hensenians.

And yet Hensen was wrong in important ways. Haeckel was guilty of fanciful theorizing, all right—and not just about plankton; he was the source, after all, of the *Bathybius*-as-globe-spanning-Urschleim theory—but theory led Hensen astray too. He started with the assumption that plankton were distributed uniformly, and then set out to prove that assumption with sampling and statistical methods that were completely inadequate to the task. Hensen could not conceive that an ocean that looked homogenous to his plankton net was anything but. Yet the plankton distribution, it is now known, really is patchy, as Haeckel, with his supposed disregard for facts, had claimed all along. Moreover, Hensen's conception of "facts" was stunted: not all important facts are quantitative ones. If you try to study the ocean exclusively with quantitative methods—without looking, as it were—you will miss some of its most fascinating phenomena. Instead of seeing the panoply of delicate gelatinous animals that dominate the open ocean, for instance, with their mysterious anatomies and behaviors, you will see only their crushed remains: a plankton net full of snot.

Biological oceanographers these days often speak of having moved "beyond mere description," of the kind Haeckel pioneered, as if quantitative knowledge were somehow higher than descriptive knowledge; as if the age of exploration had come to an end with the nineteenth century. The opposite is true: the exploration of the ocean's vast volume has barely begun. Being ourselves rooted to a surface, we naturally tend to focus on surfaces; and when we dive below the surface of the sea, we usually head straight for the next one, the seafloor. But between the top and the bottom of the ocean lies 250 million cubic miles of water, around four-fifths of it deep ocean, far from shore: blue water. All that water is inhabited, and only a tiny fraction of it has been explored with anything but nets, which generally kill the inhabitants.

There are a few Haeckelians left in biological oceanography today who are exploring that blue water with their own eyes, but by and large, in a field that has to struggle for public support and that is dominated by Hensenians, they have a hard time of it. Harbison is one of them; Hamner is another. "People used to study birds a century ago by shooting them," says Hamner, the erstwhile ornithologist. "It took people like Roger Tory Peterson to show that you didn't have to blow the birds away— to show that you can identify live birds by sight or song. But most oceanographers still only look at crushed animals. And there are some really beautiful animals in the ocean." Snails with wings, for example.

In 1971, two years after he first jumped into the Gulf of California, Hamner led an expeditionary force of young blue-water divers to Bimini, in the Bahamas. Their youth would prove to be an essential adaptation to their living conditions. Hamner himself was just 33 then but already a kind of Pied Piper. It was he who had managed to surround himself with a large group of undergraduate and graduate students who were willing to follow him, methodologically speaking, into uncharted waters. It was he who had managed to convince the Guggenheim and other foundations to fund a group sabbatical in the Bahamas, and the Boston Whaler company to provide a boat. The funds were not lavish, however, and so Hamner required anyone with an independent income to donate it to the commonweal; everyone then got an allowance of $5 a week. In a large unfinished house rented from a local minister, with plywood floors, holes for windows, private rooms only for the married couples and one bathroom for all, a shifting group of 12 to 30 biologists, would-be biologists, and children established a kind of scientific commune. For a year and a half, they lived on fish and rice. They went to their air-conditioned laboratory, which belonged to the American Museum of Natural History, to stay cool and to get away from the kids. And most of all they dove.

Bimini, perched as it is on a steep seafloor shelf, was perfect for what Hamner had in mind. Just half a mile offshore, you could be in the Gulf Stream, in water that was 2,000 feet deep—in other words, in the open ocean. You did not need a large oceanographic research vessel and a large budget to get there; you could go out in a Whaler and be there in a few minutes. This meant you could visit the creatures of the open ocean every day or even twice a day. You could watch them as a bird-watcher might, and begin to understand their behavior as individuals.

By then Hamner had devised safe diving methods. That you did not dive into blue water alone went without saying. Hamner preferred to go out with five people: one to wait in the Whaler, three to do research, and one safety diver to keep the other three out of trouble. The three research divers were tethered by 50-foot lines to an aluminum bar that was tied to the safety diver, who could thus keep his colleagues from vanishing into the abyss.

They in turn could devote themselves fully, for an hour or so, to watching animals, photographing animals, collecting them in nets and jars, and burbling observations into water-proof microphones. All they had to do was keep an eye out for sharks; a six-foot gray once rammed into Hamner's wife Peggy, also a biologist. "You learn to deal with them like you deal with big dogs in the street," Hamner says. Whenever a shark appeared, the divers would simply huddle together back to back, brandish their shark billies, and wait for it to go away. In their long stay on Bimini they were chased from the water only once, the first time they saw one.

During that time they never saw the same water twice. Off

Bimini the Florida Current, fountainhead of the Gulf Stream, moves through the Strait of Florida at around two to three knots; the drifting animals Hamner and the other divers saw one day might have moved 70 miles north by the next. Meanwhile a fresh troop of pilgrims had streamed in from the south. It was in October of the first year at Bimini that the researchers witnessed the arrival of the pteropods—the wing-footed snails.

"It was just this really strange water mass that came through for about a week," recalls Ron Gilmer, who at the undergraduate age of 20 was the team's designated mollusk expert. "We never really figured out where it was from. But it was chock-a-block full of big pteropods like *Gleba* and *Corolla*, and they were making these huge mucous webs. Some of the animals were as big as your fist and were making webs that were as big as a dining room table in the water above them. There were so many in the water that their webs were almost continuous.

"Bill was actually the first to see one. He knew it was a mollusk, but wasn't quite sure what it was. Bill said he had to really swim to catch it, that the thing was outswimming him— Bill can be a little flowery at times. So I got out there, and they really are incredibly fast swimmers! It's almost impossible to catch one if you disturb it, if you haven't drifted in real close. It's sort of like catching a fly—if you get really close to it you might be able to get it in a jar, but otherwise forget it."

Like catching a fly or perhaps a butterfly: pteropods belonging to the same family as the ones Gilmer saw in 1971 were known to Provençal fishermen as early as the eighteenth century, and dubbed *papillons de mer*, or sea butterflies, for the fluttering motion their wings make as they swim. And as early as 1761, the exact same species that Gilmer saw off Bimini, *Gleba cordata*, had been reported off Arabia by Danish explorers. (It is actually rare in the western Atlantic.) Throughout the nineteenth century a succession of well-known naturalists turned their attention to the bizarre little creatures. Baron Georges de Cuvier, having named snails "gastropods" for the foot that comes out of their stomach, coined the term "pteropods" in 1804; and in 1810 François Peron, examining samples collected on his circumnavigation aboard the *Naturaliste*, pointed out the most fundamental way of distinguishing among pteropods: some have shells and some do not.

All, however, are snails that have taken flight from their various substrates into the ecological vacancy of the open water. And until Gilmer entered that space too and got close to some living pteropods, their central mystery remained: How did they do it? How exactly did they make a living? All the dissections of dead specimens by all the learned men since Cuvier had not been able to answer that question, especially in the case of the shelled pteropods. To lug those heavy shells around, those vestiges of a former existence spent crawling, not swimming, it seemed they must flap their little wings furiously all the time. It was hard to see how they could get enough food in the desert waters of the open ocean to stoke such a busy metabolism.

The answer turned out to be fairly simple once you saw them alive. There were really two answers, Gilmer discovered, and the discovery earned him a paper in *Science* before he even had his B.S. First of all, pteropods do not have to beat their wings constantly to keep from sinking; they float. How exactly they achieve neutral buoyancy is still not clear, and in any case the mechanism seems to vary from species to species. But even the most encumbered of shelled pteropods manage to stay afloat somehow, or at least to sink extremely slowly. And while hanging there in the water—as Gilmer discovered while he hung near them—they have no trouble feeding to satiety. Their trick is to cast a wide net: those giant webs of mucus that pteropods secrete are designed to catch plankton.

Gleba cordata is an especially advanced player of the game. Though it is technically a shelled pteropod, it has a pseudoshell—a gelatinous conch that provides structural integrity at much less freight than a shell of calcium carbonate. The conch is about an inch and a half long, and proceeding from its ventral opening and extending about a conch length to either side is a single broad and muscular wing. The wing has the wavy grace of a ray's body. *Gleba* floats on its back when feeding, with the conch hanging down and the wing on top. Rising up from the wing is a slender, half-foot-long proboscis—*Gleba*'s version of a snail foot. Inside the proboscis is an esophagus, at the top of the proboscis is a mouth, and around the mouth are tiny ciliated lips. If *Gleba* were forced to collect food with those lips alone, it would starve. But spreading out from the mouth, balanced on it like a plate on a juggler's stick, is six-foot-wide sheet of sticky, free-floating mucus. The mucus collects a lot of plankton indeed—so much that when the snail periodically reels in its web, it has more food than it can cram down its esophagus and has to shed some of it in long strings.

After swallowing a web and crushing the contents in its muscular gizzard, *Gleba* promptly casts a new one, secreting the mucus from glands that rim its wings. In hours of watching, Gilmer never caught the snail in the act of casting; pteropods are skittish beasts. When you get too close to a feeding *Gleba*, it cuts loose from its web, flips over so that its wings are below the conch, and darts off at about a mile per hour. Gilmer was never able to watch an individual pteropod for more than three minutes at a time. But in a decade and a half of diving everywhere from Greenland to Antarctica, with Hamner and later with Richard Harbison, he verified that *Gleba*'s mucous web is not an idiosyncracy. All the shelled pteropods seem to feed that way, just as they all seem to float—not that you would ever know that if you only watched their constrained behavior in a laboratory aquarium. As recently as 1992 Harbison and Gilmer had to write a scathing note to the *Journal of Molluscan Studies;* some British researchers who had observed pteropods in their lab had concluded that they sink like stones when they do not flap as frantically as hummingbirds. The problem was, Harbison and Gilmer had seen the same species in the ocean floating effortlessly. The sinking ones in the lab, they said in their note, were probably just trying to get away from their tormenters.

———

Jelly, jelly everywhere: it was not just the mollusks at Bimini that were gelatinous. There were of course jellyfish properly speaking—the medusae, members of the cnidarian phylum, which also includes corals and siphonophores. Hamner focused his own attention on them. Meanwhile two of his graduate students, Alice Alldredge and Larry Madin, divided up the chordates—the phylum that includes us. We who have backbones belong to the subphylum of vertebrates, but in the subphylum of tunicates there are two classes of very common gelatinous zooplankton, the larvaceans and the salps. Alldredge and Madin flipped a coin, and she got the larvaceans. We hardly recognize them as cousins, and they are distant ones to be sure. But in one respect, at least, they are quite obviously like man: they are ambitious architects.

Whereas pteropods get their plankton with tables of mucus, larvaceans build whole houses, with rooms and halls, and live inside them. The houses are prodigious plankton nets. In fact, one of the early students of larvaceans, a turn-of-the-century German biologist named Hans Lohmann, was able to describe many new species of microscopic plankton by picking them off larvacean houses. The larvaceans themselves are tiny but not microscopic. The animal looks like a tadpole—the stiff notochord running down the center of its tail identifies it as a chordate—and is typically less than a fifth of an inch long. The house, though, can be the size of a walnut, and sometimes much bigger.

Lohmann and others after him had described larvacean houses, but Alldredge was the first to describe them and photograph them alive, in situ, and to determine in detail how they work. How they work is pretty remarkable. In the genus

Oikopleura, for instance, which Alldredge saw a lot of at Bimini, the animal sits at the center of its house with its mouth facing backward—that direction being defined by the flow of water through the house. As the animal beats its tail vigorously, water is drawn in through two openings at the front; the openings are covered by a fine mesh of mucous fibers that strain out overlarge particles that would otherwise choke the larvacean. The water then courses past the tail, into the rear part of the house, through two passageways that curve around the outer rims of a pair of arced and layered wings of mucus. Flowing between the layers and up the arcs, toward the centerline where the wings meet, the water deposits its freight of food particles; then it squirts out the back of the animal. Every few seconds the larvacean sucks all the trapped food into its mouth, by means of a concerted beating of cilia in its gills.

It is a beautiful design—but not without its flaws. One is that the external filters tend to get clogged, even when the animal beats its tail in reverse (as some larvaceans can) to flush them out. Another flaw is that the animal's anus empties waste inside the house rather than outside. For both reasons a larvacean cannot treat its house as a permanent home; it has to make a new one several times a day at least. While it is feeding tranquilly in its old home, its skin is already secreting the mucus for the new one, which lies draped over its trunk like an uninflated balloon. When the time comes, the animal inflates it by tumbling violently to shake the house open and finally by swimming inside and beating its tail. Meanwhile it has cast the foul old house aside. In areas where larvaceans are abundant, which are many areas of the ocean, the water is littered with abandoned and slowly sinking larvacean houses. Diving through them, Alldredge found, is like diving through a snowstorm.

Salps contribute to the snow of mucus too, but on the whole they are less profligate with it: they recycle most of it rather than throw it away. Salps are open-ended barrels of elastic gelatin, several inches long and bound by hoops of muscle—sleek little ramjets of jelly, they are—and for them, as Larry Madin discovered at Bimini, to move is to feed. As water flows into the mouth of the barrel, first the muscular lip and then the body of the salp contracts, shooting the water out the back and pushing the animal forward. On its way through the animal the water passes through a conical net of mucus—like the plankton net of an oceanographer but much finer—that is suspended from the inside of the barrel. The net strains particles as small as bacteria from the water.

Salps, Madin found, are pretty much always moving and always eating. Floating next to them, Madin squirted red carmine dye into the water just before it flowed into the barrels. Then he watched as the particles of dye got stuck in the net and rendered it visible—and were slowly, steadily carried back towards its apex at the esophagus. The net was like a conveyor belt: it was continuously secreted near the front of the salp and continuously devoured, along with its content of food, at the back. Madin clocked the progress of the net and found that it took anywhere from 10 seconds to four minutes, depending on the species of salp, to turn over completely. He clocked the salps themselves with a stopwatch, swimming freestyle or in plastic tubes, and found that they made around a tenth of a mile per hour, on average. Not fast, but steady and methodical—nothing much that is edible in front of it escapes a salp.

That efficient feeding mechanism is one reason why the water in the open ocean is sometimes thick with salps. Another reason is that they are highly efficient—and exotic—reproducers. A solitary salp is a population boom in the making. As it feeds, a linked chain of daughter clones—Madin likens them to paper dolls—is already budding from its back, and they too are already feeding. Eventually this asexually spawned chain breaks off from its parent and floats free. In some species it is a simple chain, in others a double helix, and in still others a wheel of salps connected by spokelike peduncles. In all cases the individual salps are destined for motherhood within a few hours, as soon as they can find a chain of fathers to fertilize them. And they themselves will have switched sexes, growing testes to become potential fathers, by the time they have given birth to a new generation of solitary salps. Sexual and asexual generations thus alternate rigorously and rapidly, marrying reproductive quality (sex mixes genes in a healthy way) with quantity (cloning cranks out lots of offspring). When conditions are right, a salp population can explode. Years after he met salps at Bimini, Madin encountered a swarm so dense off British Columbia that he figured it was clearing half the water in the area of every food particle in it—every day.

No one really knows what the right conditions are, why salp swarms occur in some times and places and not others. But one thing that seems to keep a salp population in check, paradoxically, is too much food. Harbison and Gilmer first noticed the phenomenon while diving in the 1970s and later confirmed it with laboratory experiments. When a salp encounters water that is too rich in plankton, it chokes. A bolus of food and mucus lodges in its esophagus like a poorly chewed ham sandwich, and while the animal can sometimes dislodge the obstruction by a kind of tunicate Heimlich maneuver, swimming backwards and blowing the whole mucous net out its front end, often that maneuver fails. A salp with a blocked esophagus, unable to feed though it be floating in lunch, is not a happy salp. "The few clogged salps we have seen in the field appeared to be in poor condition," Harbison and Gilmer wrote.

Clogged salps are fairly rare in the field, Harbison thinks, because salps steer clear of danger zones. With the exception of a few species, they avoid the rich coastal waters; they live far from shore, which is why they are so unfamiliar to most of us. With their delicate vacuum cleaner of a feeding apparatus, they are especially adapted to life in the open ocean. Where food is plentiful, they starve; where it is small and scarce, they thrive. Salps like it out there.

Harbison got his first taste of blue water on a transatlantic cruise in 1972, while Madin and the others were still at Bimini. He had always loved tunicates, he says—his excuse being that he grew up in Florida, where the piers are encrusted with sea squirts, the sessile cousins of salps. The line from boyhood passion to adult profession was not straight, however. In college Harbison started as an English major, soon tired of that, switched to history and Oriental civilization, and only in his senior year, in a zoology class for nonmajors, began to find his vocation. In graduate school he detoured into biochemistry, thinking he would study tunicate enzymes. But the cruise in 1972, from Newfoundland to Lisbon on a small Woods Hole research vessel, the *Chain*, was a turning point for Harbison. Steaming out of St. Johns, the ship passed through fog and icebergs on the Grand Banks, then pod after pod of humpback whales—through enough mystery and poetry, in other words, to bind an open young mind to the sea for life. And then, in warmer climes, came the midocean swim calls. Jumping off the ship, Harbison got a chance to swim with the salps. "I said to myself, 'This is a hell of a lot better than being in a lab coat,'" he recalls. "'This is really fun.'"

Harbison even got his first research paper out of that cruise. It was about a jellyfish rather than a tunicate, but a very unusual jellyfish. When the trawl came over the side with it one day off the Azores, most of the ship's company, seeing the giant purple arm sticking through the mesh, thought they had bagged *Architeuthis*, the elusive giant squid. They were somewhat disappointed to find it was merely a five-foot-wide purple medusa—*Stygiomedusa fabulosa*, to be precise. Harbison was not disappointed at all. The largest container on board was a

35-gallon drum, which was not large enough to preserve a 25-gallon jellyfish without mangling it. So Harbison cut a pie quarter out of the medusa and pickled that in formalin. Back at Woods Hole he discovered he had landed the largest-known specimen of one of the largest-known jellyfish species, which made it worth discussing in print. At the end of the paper Harbison and his co-authors thanked their shipmates for their help in "subduing" the specimen—a euphemistic evocation of a zany scene in which half a dozen grown men had wrestled with a giant gooey hockey puck as it slid around the deck of the tossing *Chain*. For years after that Harbison carried an empty 55-gallon drum with him on every cruise, just to be ready. But he did not catch another *S. fabulosa* until 1985, after he had stopped. The ocean is fluky that way.

Indeed one of the frustrating things, if you are a student of salps or of any other kind of jelly, is that you can organize a cruise and hire a ship and spend two weeks steaming around some plausible bit of ocean and still never see enough of the animals you came for to write home about. That happened to Harbison and Madin not long after Madin had moved to Woods Hole in 1974 to join Harbison in common pursuit. One time when they were pursuing salps, diving off a ship in the Caribbean, salps were not to be found. On the other hand, there were plenty of ctenophores in the water. Harbison and Madin decided they should probably start studying ctenophores as well. Soon thereafter, when they got the chance to start diving into deeper waters in submersibles, they would be glad of that decision.

Ctenophores are transparent and gelatinous like larvaceans and salps, but they belong to an entirely different phylum—their very own. Commonly known as comb jellies, they have eight rows of tiny paddles—the combs—running the length of their bodies, like the stitching on a football. Each paddle is really more like a paint brush than a comb, being made up of tens of thousands of cilia. Some ctenophores can swim by flapping the large gelatinous lobes around their mouth, and when they do that they look almost conventional by marine standards, sort of like a jellyfish or a squid. But the rows of tiny paddles are the main source of locomotion: they beat in waves that travel down each row from the back of the ctenophore to the front, like rows of falling dominoes, or like the lights on a movie marquee, if a diver happens to be illuminating them. Swimming slowly or just hovering, waves of light rippling down the seams of its diaphanous body, a ctenophore looks anything but conventional. It looks eerily gorgeous and alien.

And menacing, if you happen to be a copepod or a salp. Ctenophores are carnivores, and they prey on all manner of plant-eating plankton. Some of them just hang there in the water, trailing long, sticky tentacles—fishing, in other words. Off Brazil, Harbison and Madin saw *Lampea pancerina*, a ctenophore less than half an inch long, fishing that way for salps. Whenever a victim stuck to its tentacles, *L. pancerina* went for a ride, because a jet-propelled salp is a stronger swimmer than a ctenophore. But slowly, as it was dragged along, the ctenophore reeled itself in, attached its mouth to the body of the salp, and proceeded to engulf it. More flexible even than a boa constrictor, *L. pancerina* can swallow a salp three times its size.

Other ctenophores hunt rather than fish. *Leucothea multicornis*, which can be as much as eight inches long, swims mouth first through the water, its large lobes billowing out to the sides like the wings of a biplane. Slender ribbons of jelly inside the lobes undulate constantly, luring or even driving prey items—copepods, pteropods, and other -pods—onto the sticky inner surface of the lobes. Once a victim is stuck there, the lobe rolls up around it, squeezing it against a tentacle that then draws it into the mouth. Diving off southern California in the mid-1980s, Bill Hamner videotaped *Leucothea* for as much as 40 minutes at a stretch. He found that it foraged, if not intelligently—that would be too strong a word for an animal without a brain—then at least "optimally," as biologists like to say. Most zooplankton migrate up and down in the water column every day, swimming toward the surface to feed under cover of darkness and hiding from their predators during the day at depths of a hundred feet or more. *Leucothea* was migrating vertically too, but it was spending most of its time swimming around horizontally in layers that were rich in copepods, mopping them up with extended lobes. This behavior is hardly surprising, Hamner pointed out, until you consider that it was being performed by a brainless and supposedly aimless gelatinous drifter.

All in all they are ferocious predators, the ctenophores, and they do not spare one another: one genus, *Beroë*, feeds primarily on other ctenophores. Another genus, *Ocyropsis*, has lobes muscular enough to throttle small fish. And yet ctenophores were also, until Hamner, Harbison, and Madin started watching them, among the least known of planktonic animals. A few coastal species had been studied since the nineteenth century, by biologists like Haeckel, who leaned over the sides of rowboats and scooped them gently out of the sea. But the great diversity of ctenophores in the open ocean had been missed entirely.

The reason was simply that ctenophores are too fragile to withstand being collected in a Hensenian way. *Leucothea*, for instance, is so wispy a current can shred it, let alone a plankton net slicing through the water at high speed as it is towed by a ship. Ctenophores come up in the net, but only as the formless mystery snot that plankton biologists have always chucked over the side after picking out the hardy copepods. In 1973, just two years before Harbison and Madin started diving for ctenophores, a Dutch researcher published a survey of the species he had managed to catch with a plankton net at 43 stations across the North Atlantic: he had found a total of two. Diving in the Atlantic and catching ctenophores in jars, like fireflies, Harbison and Madin found at least 15 different species.

And it was only when they began diving deeper in submersibles that they truly entered the realm of the ctenophores. Harbison had gotten an inkling of what was down there in the early 1970s, when he tagged along on a couple of *Alvin* dives. On one of them he convinced the pilot to leave the outside lights on during the descent; usually *Alvin* saves its lights and thus its batteries until it gets to the seafloor, which is what the people inside are generally interested in. "When we went down we could see all these wonderful jellies in the water," Harbison recalls. "You could see them going by your window, and say 'Gee, I wonder what that is.' It was pretty obvious that there was a lot of stuff in the water."

In 1976 Harbison and Madin got a chance to stop and look at some of that stuff—four *Alvin* dives of their very own, on which they could hover in midwater instead of racing straight to the bottom. From a bit of dryer hose, a pump, and other scrounged equipment, they cobbled together a "slurp gun," a sort of vacuum cleaner for collecting jellies. They took their dives on the Atlantic continental slope, just a day's steam south of Woods Hole. On their very first dive, which lasted a little over an hour, they counted 93 specimens of a ctenophore neither of them had ever seen before. In fact no one had ever seen it: it was a new species in a new genus, *Bathocyroe fosteri*. It had a distinctive, bright-red gut, perhaps to hide the flashes of light given off by bioluminescent prey as they disappeared into its entrails. Madin and Harbison had to draw and photograph their specimens fast: once on board ship the ctenophores quickly disintegrated.

Yet the ease of the discovery astonished them; in their first few minutes of looking they found an abundant new ctenophore just under the surface in a part of the ocean that has been crisscrossed and trawled by research vessels for decades. *Alvin*, after all, is hardly the ideal vehicle for exploring the midwater. It has tiny portholes, and they point down, the better to explore the seafloor. In the 1980s, Harbison and Madin got a chance to make extensive use of a more suitable submersible: the *Johnson Sea Link*, operated by the Harbor Branch Oceanographic Institution in Florida. The *Sea Link* has a gentle slurp gun and other advanced collection devices. Best of all it has a cockpit like a helicopter's, an acrylic sphere that affords a panoramic view.

Looking out that picture window, Harbison and Madin discovered that they had just been scratching the surface of the ctenophore phylum. As they sank past the top few hundred feet of the ocean, ctenophores were often the most common animal in the water. And the ctenophores they saw were not like the ones near the surface. Off the Tortugas, Harbison ran into the ctenophore counterpart of the fabulous jellyfish that had started his career—a six-foot-wide monster that was completely unslurpable and could thus only be videotaped. All in all he and Madin discovered more than 40 new species of ctenophore during the 1980s, nearly doubling the known population of the phylum.

With that heady decade now over, and funds for such exploration tightened to the vanishing point, Harbison spends much of his time in his basement laboratory at Woods Hole analyzing his collections, describing, and naming. There is enough to do. "The neatest thing about ctenophores is that they're very simple animals," he says. "They're among the most simple animals in the world. They've got these eight comb rows, they've got a mouth and no anus—the undigested stuff goes out the mouth, and the metabolic waste just diffuses out—and they've got a couple of tentacles. So there's not much to them. And every time I see a new one I'm just dumbfounded that there's another way to do something with this simple plan. What you can do with eight comb rows and tentacles and a mouth—it's just astounding. They always surprise me."

In the bewildering variety of forms that exist among the large animal plankton of the open sea, there are nevertheless a few constants. One of course is the tendency to be gelatinous. That allows the animals to float, and it also makes them transparent. Why do they want to be transparent? The answer is not as obvious as you might think. One theory—advanced by Hamner and his Bimini team—is that transparency is a way of hiding from predators. Anyone who has spent time trying to watch jellies appreciates the appeal of that theory; divers find them extremely hard to see. And yet, as Harbison points out, many animals that are mostly transparent also have parts that make them downright conspicuous—think of *Bathocyroe*'s cherry-red gut, for instance. Harbison believes that gelatinous plankton are transparent primarily because there is no reason not to be. The function of pigmented skin is to protect against ultraviolet light, he says, and most gelatinous animals spend most of their time below the sea surface, where no ultraviolet light to speak of penetrates. Why waste energy on being colorful?

On the other hand, gelatinous plankton spend considerable energy making light of their own. Their bioluminescence is triggered by a chemical reaction catalyzed by an enzyme, which produces an excited molecule, which then sheds its excitement in the form of a quantum of light. Stoking that reaction takes a lot of fuel. Yet in spite of their food-poor environment, most open-ocean jellies seem to do it, sometimes spectacularly. Their motives are uncertain. Edith Widder, a biologist at Harbor Branch, has devoted her career and 15 years of her life to studying bioluminescence in the sea—a calling that places her in a very small group. She well remembers her first encounter with her subject, the one that hooked her. It happened while she was inside a large and vaguely insectlike suit of spun aluminum, which was painted black and yellow and was called *Wasp*.

People who have dived in *Wasp* do not forget the experience. Originally designed for maintenance work on offshore oil rigs, it was a cross between a tethered diving suit and a personal submarine, with huge, cumbersome arms and no legs at all—"like being a Michelin man in a garbage can," says Harbison. *Wasp* could take a diver to 2,000 feet while keeping the pressure inside at one atmosphere; it let the diver maneuver around with the help of thrusters controlled by foot pedals; but pleasant it was not. Harbison's most vivid memory of *Wasp* is his last dive, on which a thruster failed and he became so dangerously entangled in the tether that he had to be fished from the water with a crane. "It was right toward the end of the cruise, and I was fairly pleased that I didn't have to go down again," he says, "because I probably would have shat myself." Widder remembers the 40-degree cold that made her teeth chatter and her lips blue, and the aluminum arms that were so stiff at high pressure she could barely move them, even though she had lifted weights for a year in preparation. (She is 5 foot 2 inches tall.) But she also remembers the view out the plexiglass dome that served as a helmet, and the view on one dive in particular, off southern California in 1982.

"I had a light meter in the suit, and I was trying to measure the irradiance from the surface," she says. "So I was working in the dark. I was probably down around 1,600 feet. I had a little LED readout that gave me my readings inside the suit, and I had my head down and was looking at this LED. And suddenly the whole inside of the suit lit up with blue light. I thought it was an electrical spark, because we'd actually been having electrical problems with the suit, and some people had taken voltage jabs off it. But the light stayed on, and I looked up—and there was this tremendously long chain of glowing luminescence. It was a siphonophore, meters long [siphonophores are chainlike jellies; whether individual or colonial is not clear] and it was glowing so bright I could see everything inside the suit."

Diving in *Wasp* could be inspiring that way, but it was not the best way to watch animals. Aside from the creature discomforts, its great flaw was the tether, which jerked the diver up and down in the water with every wave that rocked the mother ship. This meant the diver was observing not the animals' response to their natural environment but their response to a giant, bouncing, metal insect. With some animals and some behaviors that might not make a difference, but in the case of bioluminescence it surely did. Mechanical disturbances stimulate bioluminescence; that has been known for as long as sailors have observed the glowing wakes of ships at night. Sitting in the stern cabin on the *Challenger* as she passed the Cape Verde Islands in 1872, Wyville Thomson found he could read by that glow, which is now known to be caused by the single-cell plankton called dinoflagellates. The larger jellies respond to mechanical stresses too, including the bobbing of a *Wasp*. But how much glowing do they do, Widder wondered, when they are all alone in the dark?

In 1984 she got a chance to find out, when she dove into the Monterey Canyon in a brand-new vehicle called *Deep Rover*. *Deep Rover* was invented by the same ingenious engineer who designed *Wasp*, an Englishman named Graham Hawkes, but it is far superior as a research tool. It is an untethered submarine, similar to the *Johnson Sea Link* but for one person, who sits comfortably inside a transparent acrylic sphere that is safe down to 3,300 feet. Widder took *Deep Rover* down to 2,000 feet off Monterey, and when she got there, she adjusted the ballast so the sub would float quietly at that depth. She turned the outside lights out. She covered the faint lights on her instrument panel. She became, in effect, a large gelatinous zooplankter, hanging motionless in the dark. "It was as black as anything you can imagine," she recalls. "No light from anywhere." For a full ten minutes Widder sat and stared into the dark, and she saw absolutely nothing. Not a single bioluminescent flash. Then she turned on the thrusters. As *Deep Rover* moved slowly forward, its lights still out, the jellies started bouncing off a screen that Widder had made for just that purpose, a bit of nylon mesh stretched over a hoop affixed to the submarine's bow. A medusa hit the screen and made a perfect circle of light. A siphonophore made a twisted luminescent arc. The animals had been there all along; they had just been holding back.

"What that tells you," says Widder, "is that these animals, although they all have this tremendous capacity for light production, use it very, very conservatively. Because every time they use it, they're at risk of exposing themselves to their predators." In fact, although some midwater animals no doubt light up spontaneously, to attract prey or a mate, Widder believes that most jellies glow only when they fear a predator is upon them. This is the "burglar-alarm" theory of bioluminescence: by turning on its lights, an animal may create enough of a scene to draw the attention of its predator's predator, and thereby perhaps save itself. The corrolary of the burglar-alarm theory is the minefield theory. It says the reason so many animals tend to hang motionless in the deep, even fish, is to avoid setting off light explosions that would expose them to their enemies—their predators or their prey. Life in the midwater, in this view, is a tense affair (though the denizens do not know it) in which everyone is waiting stealthily in the dark, moving slowly if at all, watching and waiting for someone to turn on a light and for something to happen.

How close are we to understanding what this life is like? Not very, says Harbison, who has spent as much time trying as anyone. We can scarcely imagine, surface dwellers that we are, the vast three-dimensional boundlessness of the open sea, in which gravity loses its importance and the only surfaces are other animals. Nothing in outer space could be more alien to us than that. And above all, says Harbison, we cannot comprehend a place so utterly permeated with randomness, a place in which there really are no fixed places at all. "When you jump in the ocean, you never know what you're going to see," he says. "Because the water is not the same water you jumped in yesterday. And for the animals that are vertically migrating, they never go up into the same water they were in the day before. Living in the ocean, you don't have the predictability that you do on land. On land you live in your house, you know what it looks like when you walk out the door—you're always in the same place. We have a very good concept of place, and where we are makes a lot of difference to us. And it does to most animals that are living in the terrestrial environment.

"Now if you're a planktonic animal—your environment is constantly new. You never know when you're going to get a meal, you just have to be prepared to take advantage of it when a meal comes along. And you never know what you're going to find. It's as if—imagine you open your door tomorrow, and out there is a big polar bear. And the next day you open your door, and out in the front yard is a Bengal tiger. And the next day there's an anteater sitting out there. I mean it's that kind of environment, which is really hard for us to understand. It is so weird we can't even begin to think about it at this stage."

A lot of different animals live in that environment, shrimp and whales and fish and squid, for instance. But none of them, Harbison says, are as perfect an expression of it as the jellies. Floating is the most sensible thing to do in a truly three-dimensional world, and being 95 percent water and gelatinous is the best way to float. It is also the best way to take advantage of the absence of light, of gravity, and of other mechanical stresses; in the slow-moving water beneath the waves, you do not need a backbone or even a shell to retain your integrity. You can cast large and delicate nets to collect the widely scattered particles of food—nets made of sticky gelatin or mucus. All those advantages explain why so many phyla have independently evolved gelatinous representatives in the open ocean, and why, conversely, you do not see many gelatinous species on land, or even in freshwater. The open ocean is their turf. Indeed, some organisms that are not gelatinous have been able to colonize the midwater only by depending on jellies.

Larry Madin noticed the phenomenon on his very first dives into blue water. Staring at salps off Bimini, he saw they were often carrying minute crustaceans. Usually those crustaceans were amphipods, members of an extremely numerous order that an untrained eye might describe as shrimplike, except that shrimp belong to the decapods, another order of crustacean. Diving around the Atlantic in the mid-1970s, Madin and Harbison discovered 18 species of amphipods, all belonging to a group called the hyperiids, that were associated with salps in various ways. A female amphipod of the genus *Phronima* eats the guts out of a salp, moves into the empty barrel, and lays her eggs there; she keeps the larvae inside until they are mature enough, then kicks them out. Adult *Lycaea* also roam around the inside of a salp, nibbling on the body wall, perhaps scarfing an embryo or two; as long as they do not sever blood vessels or consume vital internal organs, the salp may survive. Madin and Harbison saw salps with large holes in their walls that were still swimming happily. But the most peaceful coexistence between crustacean parasite and salp is achieved by the amphipod *Vibilia*. Adults of that genus perch on the salp's gill bar, where the animal's mucous feeding net is rolled into a fine strand and funneled into the esophagus; sticking their heads right down the esophagus and grasping the strand with tiny feet, the amphipods divert some it into their own greedy mouths. The salp does not seem to mind, but in any case it is defenseless.

Amphipods invade other jellies too, burrowing into siphonophores or medusae, for instance. And amphipods are not the only kind of parasite. Harbison and Madin have seen crabs riding on salps and fish inside them. They have seen lobster larvae and shrimp hanging onto medusae. They have seen worms inside ctenophores. Other researchers have documented how copepods feed on larvacean houses, either live or abandoned ones. Many of these transactions involve food, and some of them perhaps transport. All of them involve place—a place to hide, to molt, to mate, or to rear young. At the most basic level, that is what jellies offer: places, islands in space, for animals that are not as well adapted as the jellies themselves are to a placeless world.

————

The jellies touch our lives too, and not just when they are washing up on the beach by the smelly truckload, or otherwise making the water unswimmable. They do that too, of course. Box jellyfish, also known as sea wasps because of their extraordinarily toxic sting, have killed more than 60 people on the north coast of Australia in the last century or so. A different and less lethal species makes swimming in Chesapeake Bay difficult every summer.

But jellies touch us more indirectly than with their nematocysts (as the stinging cells are called). They affect Earth's climate, for instance. The connection goes like this: carbon in the atmosphere, in the form of carbon dioxide, warms the planet through the greenhouse effect; conversely, taking carbon out of the atmosphere and depositing it in deep-sea sediments cools the planet. The mucus shed by gelatinous animals such as larvaceans carries a lot of carbon into the deep—both in the mucus itself and in the microscopic plants, animals, and bacteria that are stuck to it. Along with other organic gunk, such as clumps of dead phytoplankton, this mucus makes up marine snow. Because of its relevance to climate, marine snow is in favor with the agencies that fund oceanographic research. Alice Alldredge pioneered the study of larvacean biology and behavior, and her site on the World Wide Web still shows a photo of her scuba diving at Bimini more than a quarter-century ago. These days, though, her research is focused more on marine snow.

Larry Madin is also now doing research related to climate—climate and fish. On Georges Bank, the main New England fishing grounds off Cape Cod, a major study is underway to try to determine how various forces, especially man-made climate change, might affect the stocks of cod and haddock. Called GLOBEC, for Global Ocean Ecosystem Dynamics, the study is complicated by the fact that man has recently decimated those fish stocks, by catching most of the fish. Nevertheless, some 125 scientists are involved in GLOBEC, and Madin is one of them. His goal is to determine the extent to which jellies on Georges Bank eat cod larvae and the things cod eat, mostly copepods. Both effects could in principle drive down the cod stock. It has been nearly a decade since Madin has been able to get government support for the sort of basic exploration of the gelatinous world he used to do, beginning at Bimini.

In the Black Sea a jelly has already had a demonstrable effect on commercial fish stocks, and it has been disastrous. The agent of disaster has been a one- to three-inch-long ctenophore called *Mnemiopsis*—an American native that is common in estuaries up and down the Atlantic coast from Woods Hole to Patagonia. Around 1982 *Mnemiopsis* parachuted into the Black Sea, presumably in the ballast water of a Russian freighter that had just returned from Havana or some other port in the Americas. It evidently found the new environment to its liking. The Black Sea was freshwater until about 12,000 years ago, when the Mediterranean broke through at the Bosporus, and it is still only half as salty—about like an estuary, in other words. An ecosystem in transition, it has a relatively limited number of indigenous animal species. In particular, there does not seem to be anything there that likes eating *Mnemiopsis* much.

By 1988 the American invader had spread all over the Black Sea, and in some patches Russian biologists were finding hundreds of ctenophores per cubic meter of water. The ctenophores were eating a lot: tiny crustaceans and other zooplankton, fish larvae—anything they could get their lobes on. Ctenophores like *Mnemiopsis* keep eating even when their guts are full; they just regurgitate the undigested plankton, having spoiled it for the other animals in the water. Not surprisingly, with a gluttonous predator multiplying out of control, stealing their food and eating their young, the Black Sea fish suffered. Between 1988 and 1990 the anchovy catch by the Russian and Turkish fleets fell by two-thirds. The catch of tylka, a herringlike fish, dropped to zero. The cost to the local economies has been put at hundreds of millions of dollars. Lately the Black Sea fish stocks have begun to recover, but only because Russian fishing activity has plummeted since the collapse of the Soviet Union.

Mnemiopsis, however, has broken out into the Mediterranean, and Harbison predicts it will spread all over there too. With the help of new seedings by other freighters, it could spread all over the world; it has already been sighted off India. Harbison got involved in the Black Sea problem through a Russian friend, and he has urged the Russians to consider introducing a commercially valuable American fish into the Black Sea to control the American ctenophores and convert some of their biomass into a form people can eat. At Woods Hole he has been experimenting with butterfish, which seem to have a taste for *Mnemiopsis*. "I really believe in it," he says. "I think it's important to try, because it's a chance to do some good in the world. Most of what we do is pretty academic— it's like abstract art or something. This is a chance to do a little bit of good. It may be the only chance I get."

But in fact there is not much money available just now for that type of research, Russian science being in a state of collapse, like Russian fishing, and the Black Sea not being an American concern. Nor is there much money for Harbison's more academic work, if the exploration of the ocean can properly be called academic. In that respect Harbison is in the same boat as Madin or just about anyone else who shares their interest in the behavior of animals in the open ocean, particularly in the midwater. Edith Widder, for example: her early career was funded by the U.S. Navy, which saw bioluminescent animals as sentries that could signal the passage of Soviet submarines; but since 1991, with the Cold War over and a fancy light meter of Widder's invention already distributed in the fleet, she has had no money to study bioluminescence. Instead she studies the vertical migration of plankton, a phenomenon of closer relevance to fisheries and climate and thus of greater interest to grant givers.

In the century-long rivalry between the intellectual descendants of Victor Hensen and those of Ernst Haeckel, the Hensenians are ascendant these days. Three decades after Bill Hamner first had the idea of jumping off a ship, a number of research groups around the world have followed his lead and are using his scuba techniques, mostly near shore. But animals in the open ocean are still rarely studied in the way he pioneered; they are rarely observed in their native habitat, and rarely do the observers have the luxury of focusing on the animals themselves rather than on their effect on fish or their role in the carbon cycle. The midwater, below the range of scuba, is visited on a regular basis only at Monterey Bay in California, where the computer mogul David Packard founded and funded the Monterey Bay Aquarium Research Institute and equipped it with a deep-diving robot. A video made from *Deep Rover*, by a biologist named Bruce Robison, had helped convince Packard that the depths of the bay were worth looking at. Robison had a long track record as an explorer: he had been one of the men who first jumped into the Gulf of California with Hamner, back in 1969, and later he had organized not only the *Deep Rover* work off Monterey but also (with Alice Alldredge) the *Wasp* project in the Santa Barbara Channel. Now, thanks to Packard's largesse, he is one of the few midwater biologists with regular access to his subject matter.

But even Robison does not get to see below about 3,000 feet. That realm is not being visited at all, except by people on their way to the seafloor, passing through in submersibles like *Alvin*, with the lights out. The two *Johnson Sea Links*, which are ideal for the purpose of studying midwater biology, are hardly ever used for that and quite often sit idle, because biologists cannot get the money to pay for using them. We live at a time when the outside chance of finding a fossil bacterium on Mars is enough to generate tremendous enthusiasm—justifiable, to be sure—for billion-dollar missions of exploration to that planet. And yet we are content to pass over in complacency and almost total ignorance the largest and strangest habitat on Earth. It is odd. It is silly even.

"It's sort of frustrating," says Harbison. "Because there's this wonderful, wonderful world down there, and it's just so hard to get to. It's an inaccessible place nobody ever goes to—so nobody cares, because nobody knows anything about it. It's kind of depressing.

"Going down and *looking* at jellies—it's really, really basic when you think about it. But this whole current of erroneous thinking in oceanography has continued for so many years. Back in the days before the *Challenger*, scientists thought that the ocean below the sunlit zone was essentially lifeless. Then some bottom trawls showed there were lots of animals living on the bottom at great depth. And people said, 'OK, that's fine, the food's all produced at the surface, but maybe some of it rains down, and these animals on the bottom can live out a nasty existence—but there's nothing in between.' And then some people invented opening-and-closing nets, and they found that's not true. There were things down there.

"Then the idea was basically, 'Well, whatever's down there, we can pretty much see it with nets.' Even today, the types of questions you hear oceanographers asking about the animals that live out there are questions like 'Where are they? How much carbon do they have in their bodies?' They're things you can do with nets, and that's it. They're not anything that's relevant to the lives of the animals, like 'What eats them? What do they eat? How do they interact, how do they get together to mate in this vast ocean?' If you look at any of the animals that live out in the open ocean and ask what's known about the way they live, it's zero.

"What we need is the same thing they had in the 18th and 19th centuries: exploration. It's a dirty word now. We're supposed to be hypothesis-testing. 'We're modern scientists, there's no exploring left to do'—that's the attitude. But I'm convinced that there's so much that we don't know about what's on this planet—when we do know about it, it will change the way we think of the world."

CHAPTER 7

Invisible
Garden

At the tip of a rocky pine-wooded peninsula, next to a dock that juts out into the winter-gray chop of West Boothbay Harbor, Maine, there is an unusual botanical garden. It belongs to the Bigelow Laboratory of Ocean Sciences, and it is housed in a small clapboard outbuilding in back of the main lab. In the basement of that building stand several aisles of glass-doored refrigerators filled with rack after rack of test tubes and flasks. The tubes contain water, and in some of them it is clear; but in most it is slightly clouded by drifting specks. Those are the plants. Some of them are green.

The plants are single living cells, and the largest is a few thousandths of an inch across. Their natural home is the top of the ocean, as far down as sunlight will penetrate. They are called phytoplankton—*phyto* from the Greek for "plant" and *plankton* from the Greek for "drifting," which is what these plants do on currents and tides. Their inconspicuous appearance and behavior belie their importance. All animal life in the sea, from copepods to jellies, from krill to whales, depends on phytoplankton. It is they that, through the process of photosynthesis, harness the sun's energy to convert carbon dioxide and water into carbohydrates—the food of life—and also into oxygen. Even the denizens of seafloor hot springs depend for their oxygen on phytoplankton at the sea surface. It is phytoplankton that gave Earth its oxygen atmosphere in the first place.

Ask an oceanographer how many species of phytoplankton there are and you will get an answer like this: "In the Gulf of Maine, in August, somewhere between 500 and 5,000." In other words, no one really knows. Researchers have been cataloging this diversity for more than a century, at least since the *Challenger* and since Victor Hensen's expedition. But just in the past two decades, as they have zoomed in on the microworld with new types of microscopes and measurement devices, they have discovered whole classes of phytoplankton they never knew existed—cells so tiny they were long mistaken for dirt. "It was just a mind-set," explains Sallie Chisholm, a biological oceanographer at MIT. "We didn't think plants could be so small." Or so abundant: In 1988 Chisholm and her colleagues discovered a new species, about 30 millionths of an inch across, that populates the ocean in concentrations as high as five million cells per ounce.

In such numbers lies ecological strength and—what is of most concern these days—an immense impact on Earth's climate. While plankton ecologists have been finding new complexity at the microscale, satellite pictures taken over the past two decades have shown a different view: carpets of phytoplankton that span oceans. Each year these huge masses of plants, along with forests on land, draw out of the atmosphere about half the carbon dioxide that we put in—carbon dioxide that would otherwise warm the planet through the greenhouse effect. In short, the phytoplankton protect us from ourselves. What we do not know is whether they will continue to do so as we continue to pollute the planet. "People have no idea what's going on out there," says Chisholm, "and they want to know what's going to happen if the ocean surface warms by three degrees. I know it's going to change the structure of the ecological community, but they want to know how. And I say 'How can I tell you how when I don't even know what's out there?'"

The collection at Bigelow is a small sample of what is out there. It includes roughly 1,500 species, most of them weeds: the species that are hard to kill. As a rule, phytoplankton are more finicky than, say, tomatoes. They have the same basic needs— nitrates and phosphates as nutrients and sunlight for energy— but each species also has more subtle and individual requirements. Tiny amounts of such metals as nickel, for instance, are essential for their growth, but a hair too much can kill them. The first time you try to grow a new species, says Bob Guillard, one of the researchers who built up the Bigelow collection, "you put it in the tube, it divides a few times [phytoplankton generally reproduce asexually, by cell division] and then the little bastards croak."

In the wild, though, phytoplankton divide madly. Like plants on land, they bloom in the spring. During the winter, the turbulence of the ocean carries them to great depth, out of reach of the sun; some species seem to hibernate, shutting down their metabolism entirely. But when the spring sun creates a surface layer of warm water that is too light to sink, the plankters come alive. In that layer everything comes together for them: suspended in sunlight, they wallow in nutrients brought up from the deep by the winter's turbulence. One by one, the populations of different species explode in an orgy of cell division, doubling as often as several times a day.

In a given region the sequence of plankton blooms is usually the same. In the Gulf of Maine, for example, the first to bloom, in April, are the diatoms—greenish-brown cells encased in transparent silica boxes decorated with intricate patterns of slits and pores. After a few weeks come the dinoflagellates, with their whiplike flagella; those propellers allow some dinoflagellates to commute back and forth every day between the surface, where they soak in sunlight, and a depth of 40 feet, where they spend the night scavenging nutrients missed by the diatoms. Last to bloom, and weirdest looking, are the coccolithophores: little cells armored in ornate shields of chalk. In June parts of the Gulf of Maine turn milky white with these coccoliths, which the coccolithophores, for reasons unknown, shed by the zillion.

Before the satellite pictures started coming in, in the late 1970s, plankton blooms were generally assumed to be confined to coastal waters, where rivers and upwelling currents channel nutrients to the surface and where profitable fisheries attest to the abundance of fish food. The satellite images changed all that. They showed, for example, that in spring the entire North Atlantic goes green with chlorophyll—the molecule that plants use to capture sunlight. The bloom starts in the south in March, along a line that stretches from North Carolina to Spain, and then sweeps north in a wave that parallels the wave of blossoming on land. Oceanographers are now getting their best view ever of that pattern thanks to an American satellite launched in the summer of 1997. Called SeaWiFS (for Sea-viewing Wide Field-of-view Sensor), it is circling the planet in a near-polar orbit, mapping the color of the entire world ocean—insofar as it is not masked by clouds—every two days. The color of the ocean reveals how much chlorophyll and other plant pigments are in the water, and thus how much plant matter. SeaWiFS will be able to show how the plant mass changes from year to year, from season to season, and even from week to week.

But even SeaWiFS misses a lot of what is happening in the sea. Satellite pictures show only plant life at the surface, whereas in fact the sunlit habitat of phytoplankton extends as much as 700 feet below the waves. As it happens, some of the plankters that are most important at depth are the very ones that microscopists also missed for a century—until the late 1970s, when oceanographers discovered the red specks that now dwell in several test tubes at Bigelow. Actually, Bob Guillard had first pulled those specks out of the ocean as early as 1965, off the coast of Surinam, but at the time he did not think much of his find. He identified the organism as a new species of, curiously enough, blue-green alga, better known now as cyanobacteria; it was red only because it contained an unusual pigment besides the usual chlorophyll. "I kept it in culture because it was pretty," Guillard recalls.

Fifteen years later, though, his red specks had become more than just another pretty plankter. By then oceanographers were taking a new tool, the fluorescence microscope, to sea with them and were coming back with startling reports. Under the blue light of the microscope, swarms of tiny motes that looked like bits of sediment or feces lit up brightly. Amazingly, they turned out to be living cells, full of photosynthetic pigments. And one of the pigments was an unusual one that caused the cells to fluoresce a striking orange.

That pigment tipped Guillard off. The glowing motes, he realized, were the same as the red specks in his test tube, only now they were all over the ocean, and in fantastic numbers: there were thousands of them in a milliliter, a cube of water one centimeter on a side. "Look," Guillard remembers saying as he thrust a flask into the hands of his colleague John Waterbury, a Woods Hole oceanographer who had done the microscopy work, "John, look—we're ass-deep in these things!" Ass-deep, that is, in a single-cell organism, about a micrometer in diameter, that could not be seen with the naked eye until millions of them were grown in a flask.

The cells of this new organism, called *Synechococcus*, were different from those of green-algae phytoplankton and all higher plants on land. *Synechococcus* and other cyanobacteria are evolutionary intermediates between plants and bacteria. They photosynthesize as plants do, but like bacteria they lack the internal cell structure of the higher plants: the nucleus containing the plant's genetic material, for example, and the chloroplast housing the photosynthetic apparatus. In fact, the higher plants are believed to have evolved by absorbing cyanobacteria and putting them to work as chloroplasts. *Synecho-*

coccus is essentially a naked chloroplast, a little photosynthesis machine. With its discovery, plankton researchers had discovered a whole new realm of their ignorance: an entirely new type of phytoplankter, incredibly abundant, that played an unknown role in the marine food web and had an unknown effect on the global environment. "A hundred years of oceanography," says Guillard, "and the most abundant being in the world was not recognized by anybody."

Only now it is no longer the most abundant. The current recordholder is the phytoplankter that Chisholm, Rob Olson of Woods Hole, and Waterbury stumbled on in the late 1980s, while they were studying the distribution of *Synechococcus*. The instrument of discovery this time was not the fluorescence microscope; in Chisholm's lab at MIT, the microscopes stand off to one side, by the window. ("We don't use them much," she says, "only to get inspired.") She and other researchers use a newer tool, called the flow cytometer, that allows them to study large numbers of individual plankters that are almost too small to look at.

In a flow cytometer, a seawater sample containing a variety of cells is compressed into a thin stream, and the cells are marched in single file, 2,000 per second, past an "interrogation point." There they are bathed in blue laser light, which causes them to fluoresce. The color of each cell's fluorescence indicates what pigments it contains; the way it scatters the laser light reveals its size and shape. The cells can then be separated into species, much as you might distinguish Japanese from Swedes, say, without looking at them, on the basis of their size and hair color. Similarly, if you were ever sent data on a population of human beings, all of whom had flaming red hair and none of whom were taller than four feet, you would know you had discovered a new people.

That is how Chisholm and Olson discovered their new species of phytoplankon. At a depth of several hundred feet, at the murky limit of the sunlit zone, they found a population of cells that were even smaller than *Synechococcus*, about 0.8 micrometer in diameter, and that also lacked a nucleus and a chloroplast. But these cells did not fluoresce orange the way *Synechococcus* does. Surprisingly, they contained chlorophyll b, a secondary type of chlorophyll that is found in all green algae and higher plants—but not in other cyanobacteria. At first Chisholm and her colleagues thought they might have found the primordial plant: the cyanobacterium that had been absorbed into a larger cell and enslaved as a chloroplast. But when they were finally able to grow their new plankter in culture and look at its genes, they discovered that evolution had played a trick on them. Although it had the same chlorophylls as chloroplasts, this plankter was not their ancestor. Chlorophyll b has apparently evolved several times independently.

Chisholm named the new species *Prochlorococcus*, meaning "little green spheres." Little as they are, they are not likely to be the last frontier in plankton discoveries. Everyone who has been to sea knows that frontier is always receding. "The diversity is incredible," says Chisholm. "There are thousands of different forms. And yet they all essentially do the same thing: they use light and water and carbon dioxide to make organic matter."

That, in a nutshell, is what the ecologist G. Evelyn Hutchinson meant when he spoke of the "paradox of the plankton," in a famous essay written in 1961. The ocean looks like a fairly homogeneous place; it is all water, after all, and fairly well mixed water at that. Yet in any given region of it there may be thousands of different species of single cell plants. How come?

The answer may be that seawater, like the seafloor, only seems to be homogeneous. When you look at it closely it turns out to be patchy, like the rest of the world. The correct scale at which to study a phytoplankter, says Farooq Azam, a microbial ecologist at the Scripps Institution, is a microliter—a cube of water one millimeter on a side. To think his way into this microworld, Azam once converted his office into a scale model. In one microliter of ocean, there would typically be a single largish phytoplankter, a diatom, say. In Azam's office it was represented by a piece of green foam the size of a baseball that hung from the ceiling. There would also be between one and five protozoa—single-cell, bacteria-gobbling animals—which in the office were orange-foam spheres as big as golf balls. Hovering among these behemoths were bacteria (bits of Styrofoam packing material) and viruses (colored pushpins). There ought to be a thousand of each, but Azam was not compulsively accurate about his model. He also did not bother to include the 10 *Synechococcus* or the 100 *Prochlorococcus* that should have been there.

Even if he had, the organisms would not have been packed shoulder to shoulder; the airspace in Azam's office was still mostly empty. On the other hand, the organisms were not far from one another when you consider their ability to move around. And according to Azam, the bacteria and the protozoa move around quite a bit: they all come together in the center of the room, forming a cluster of life that revolves around the phytoplankter. The phytoplankter is the nucleus because it is the one organism that can turn carbon dioxide into new organic matter. But to do that it needs a continuous supply of nutrients, which it gets from the bacteria and the protozoa. The plankter extrudes a sticky strip of polysaccharide onto its surface that draws other organic detritus—loose proteins and DNA—out of the water. This mixture is the daily bread of bacteria, and they head for it. More precisely, they run and tumble, whirling their flagella first one way, then the other, and stopping after each cycle to measure whether the resulting zigzag is getting them closer to the food. Surprisingly, they reach the phytoplankter in minutes—in spite of the fact that, to a 0.2-micrometer bacterium, the water feels as thick as honey.

Once arrived, the bacteria feast on the organic stuff on the plankter's surface, breaking it down with enzymes and absorbing the pieces through their membranes. But in the process they throw away ammonia (a compound of nitrogen and hydrogen) and phosphates—the very things the plankter needs to synthesize new organic compounds. By simply extruding a few polysaccharides as bait, the plankter has managed to surround itself with its own personal cloud of food. The concentration of nutrients nearby might be quite small—in fact an oceanographer who sampled a whole quart of the water might conclude it was a desert—but the plankter is provided for. So are the protozoa: they cluster around the plankter, nibbling at the bacteria and perhaps at the *Synechococcus* and *Prochlorococcus*. In the process they too release dissolved nutrients into the water like so many table scraps.

At any moment, of course, this idyll may be shattered when some terrifyingly large multicellular animal—a two-millimeter-long copepod, say, which at the scale of the model would fill Azam's office—comes crashing through and devours the diatom. But often as not that does not happen. Often as not, organic matter in the ocean does not flow, as it used to be thought, in a straight chain from phytoplankton to multicellular zooplankton to fish. That may be true in the nutrient-rich waters near the coasts, but in the open sea, it is now clear, much of the organic matter gets caught in the "microbial loop" of phytoplankton, bacteria, and protozoa. And when the phytoplankter dies, the cycling and recycling does not end right away. The plankter sinks, slowly at first, until it sticks to others of its kind or to a flake of heavy mucus sloughed off by one of the numerous gelatinous animals in the sea—that is, to a flake of marine snow. Then it rides that fast-sinking particle into the abyss. Even in death, though, it continues to be colonized by bacteria and protozoa. Almost desperately, they try to feed on its cache of organic matter before it is lost to them forever.

Strange as it may sound, we are all affected by this microcosmic struggle—because it is repeated every minute, throughout the world ocean, an astronomical number of times. The phytoplankter makes its organic matter from dissolved carbon dioxide, which is ultimately drawn out of the atmosphere. When it dies and sinks into the abyss, its carbon is removed from contact with the atmosphere for hundreds of years, until ocean currents bring the carbon back up. In other words, the phytoplankton, together with marine snow, act as a biological pump that transports carbon from the atmosphere to the deep ocean. But the bacteria and protozoa short-circuit the pump. In breaking down the organic matter, they respire carbon dioxide just like we do and return it to the ocean surface and thence to the atmosphere. How well the phytoplankton do their job of fixing carbon, and the bacteria and protozoa their job of releasing it, determines how much carbon dioxide comes back into our air.

As it turns out, there are large regions of the ocean—the northern Pacific off Alaska, the equatorial Pacific, and the entire Southern Ocean around Antarctica, south of a latitude of 40 degrees—where the phytoplankton do not do as good a job as they might. In those vast regions there should by rights be rain forests of plankton, and instead the water is closer to a desert. If ecologists on land could not explain why there should be deserts in one place and rain forests in another, it would suggest a fundamental gap in their understanding, and yet for decades that was the situation in biological oceanography. But in the late 1980s and early 1990s the mystery was solved, thanks above all to a man named John Martin.

––––––––

At the time of his death from prostate cancer in 1993 at age 58, Martin was the director of the Moss Landing Marine Laboratories, a small research center on Monterey Bay. He had started out as a plankton biologist, leaving the University of

Rhode Island in 1966 with a Ph.D. in oceanography, but had then taken a career-long detour into trace-metal chemistry. His first job was with the Atomic Energy Commission, which sent him to Panama to work on Project Plowshare. It seems plausible, though Martin himself might have disputed it, that his early exposure to that ambitious effort conditioned his mind somewhat; so that when he later suggested that dumping 300,000 tons of iron in the Southern Ocean might be a good way to combat the greenhouse effect—because it would make the plankton there bloom as they had not bloomed since the last ice age, because it was iron that the plankton lacked—the idea did not seem as absurd and evil to Martin as it did to many of his peers. Compared with Project Plowshare, fertilizing the Southern Ocean seems almost temperate. The goal of the AEC project, after all, had been to plow a second, sea-level canal through the Panamanian isthmus by detonating a chain of 150 or so atom bombs.

"What you do is you set the devices up across Panama, and you touch them off, and it's just like a wave in the ground," Martin recalled years later, in 1990. "You end up with a depression half a mile wide and banks on either side. But when you blow off all these underground shots, some radioactivity is going to fall out on the sea surface. And so they wanted to find out how much of this was going to be taken up by the food chain and eventually get back to man and become a human health hazard. In the old days of atomic testing they used the 'specific activity' approach, which said the more of a stable element there was to dilute the radioactive element, the safer you are. In other words, if you're a sea creature and you've got an atom of radioactive zinc there in the water, and a hundred atoms of stable zinc, the chances will be a hundred to one that you'll take up the stable element. So they wanted to know how much stable zinc was in the water, and iron and all these other things.

"Anyway, the bottom line was you have a big mountain

range in Panama, and when you knocked down the mountain it would have blown out every window in Panama City. So that kind of slowed them down. But it was a very interesting project, and a tremendous amount of research came out of it."

Martin's own research involved taking samples of plankton all over the Gulf of Panama and measuring their metal content. He found that the water and the plankton contained quite a bit of metals already, which suggested to the AEC that a little radioactive fallout would not do much harm. Martin got pretty good at measuring metals, or so he thought at the time. But by 1970, when he took a job at the Hopkins Marine Station, a Stanford University outpost on the Monterey peninsula, he was tired of plankton ("They're cruddy things," he explained) and he was tired of metals. He wanted to switch to something new.

A colleague pointed out that his timing was all wrong. The early 1970s were a time of swelling popular concern about the effects of trace metals in the environment. Caltech geochemist Clair Patterson, in particular, was warning about the pervasiveness of lead pollution and about the dangers of lead poisoning. The Roman empire had fallen, Patterson said, not to the Visigoths and the Huns but to lead poisoning. Prodded by his colleague at Hopkins, Martin read Patterson's papers. They convinced him he should stay in the trace-metal business. When he moved to Moss Landing in 1972, he set up a trace-metal lab.

But Patterson proved to be a difficult role model. "About the mid-70s," Martin said, "Patterson started telling everybody, 'Your numbers are all bad. You're contaminating your samples. When you try to measure lead in water or in plankton, everywhere you look or touch is contaminated with lead that's flowing out of exhaust pipes, and if you could put on this special pair of glasses, you'd see lead all over the room.' And Patterson was enough of a strong individual to go into a room full of scientists and say, 'You're all full of shit!' Then everybody would scream at him, 'You stupid old man!' He just completely infuriated a lot of people, and so they didn't want to believe him."

Martin and his assistants, though, decided to heed Patterson's advice. They started taking off their shoes before entering the laboratory so they would not track in metal-rich dirt. They wrapped each bottle of seawater in three plastic bags. They kept the laboratory under positive pressure, with filtered air coming in and fans blowing out, so that no dust grains would waft in when the door was opened. They even switched to the brand of laboratory tissue that Patterson recommended as being particularly metal-free. "Our big goal in life," said Martin, "was to get one goddamn number for lead that Patterson would believe—in plankton, in sea lions, in water, or whatever. So we started doing all this, and it was amazing. All of sudden all these things started to fall into place."

In Panama Martin had found that many metals were naturally abundant in seawater. Now he found that he had been wrong. When he cleaned up his procedures and elimated sources of contamination, the water turned out to be a lot purer. Moreover, such metals as were in it were distributed in an orderly way. Before, when Martin and other researchers took samples of the water column at progressively greater depths, the concentration of a metal would fluctuate all over the place, inexplicably. Now it followed a smooth profile, generally increasing with proximity to the metal-rich mud on the bottom. Things made sense.

Martin and his group measured lots of different metals: mercury, cadmium, manganese, lead, zinc. But the biggest challenge was iron. Martin had measured it in Panama too, but those numbers, he now realized, were far too large. Iron is highly insoluble in seawater; it tends to glom onto organic particles and sink to the bottom. So the amount of iron in the water column had to be very small. But nobody had been able to say exactly how small, because we live, as it were, in a cloud of the stuff. Iron is the fourth most abundant element in

Earth's crust, which makes contamination almost impossible to avoid. "Everytime you get a speck of dirt, that's five percent iron," said Martin, "whereas it's only got a trace of zinc. And of course man uses iron for everything. When you make a plastic bottle it's usually extruded onto a stainless steel form—there's this hot plastic flowing over stainless steel, and some of that stainless steel is picked up in the plastic." Even the plastic bottles in the laboratory, then, were not to be trusted as far as iron was concerned.

But after 10 years of being washed in nitric and hydrochloric acid—which is the procedure that Martin's team followed with all the implements and even the walls in their all-plastic laboratory—a filthy, iron-infested bottle can get pretty clean. In a nutshell, that is the secret of how Martin's colleague Michael Gordon finally succeeded, in the early 1980s, in measuring the first reliable profile of iron in seawater: he finally had a clean enough bottle. By then Martin had retired from lab work. Reluctantly, he had also quit going to sea. Over the years, the polio he contracted in college in 1954 had made him progressively less mobile, and finally it had confined him to a desk and a wheelchair. Gordon and another colleague, Steve Fitzwater, had taken over the hands-on work. It is slow, craftsmanlike work, and it requires dexterity. It is work that can be completely undone by a single errant dust grain. Gordon's iron profile, taken on a cruise in the North Pacific, showed that the concentration of iron in the surface waters, at least in that part of the ocean, was ridiculously small: a few parts per trillion, a few micrograms per cubic meter, or less than a millionth of an ounce per ton.

When Gordon showed Martin this result, Martin at first did not pay much attention; it was a remarkable measurement, sure, but it lacked a context that would have made it interesting. But then at a meeting of plankton mavens in December 1986 Martin heard a talk by Bruce Frost of the University of Washington. It was about that old puzzle in biological oceanography, the "high-nutrient/low-chlorophyll" regions of the ocean. In all three of those regions, the northern and the equatorial Pacific and the Southern Ocean, Frost explained, upwelling currents from the deep ocean supply phytoplankton with the main nutrients—the nitrates and phosphates—they need to grow. Yet a lot of the nutrients go unused. The phytoplankton never convert them into chlorophyll, and never bloom as they do in the North Atlantic and other regions. Why? Frost offered the conventional explanations. Maybe there is not enough light, he said, or maybe it is too cold, especially in the Antarctic. Or maybe the phytoplankton are kept in check by vigorously grazing zooplankton—in other words, maybe the grass is short because of the presence of sheep. As Martin listened to all this, a different idea came to him. After the talk he went up to see Frost. "Aw baloney," he said. "It's just iron deficiency."

Like every other living organism, phytoplankton need iron. They need it in particular to make chlorophyll for catching sunlight; to make nitrate reductase, the enzyme with which they reduce nitrate to a form suitable for making proteins; and also to make DNA. So in principle it was possible for a phytoplankton population to be hobbled by a lack of iron—but in practice it was a little hard to imagine, because they need so little of the stuff. Just how much they need is not known precisely even now, and it may vary from species to species. But the best estimate available when Martin made his half-joking remark to Frost was that the typical planter uses at most 1 atom of iron for every 10,000 atoms of carbon, 1,500 atoms of nitrogen, and 100 atoms of phosphorus it assembles into protein, DNA, and the like. That is not a lot of iron.

And yet, when Martin started trying to turn his quip into a serious hypothesis, he soon found evidence that phytoplankton really might not be getting enough iron, at least not in the high-nutrient/low-chlorophyll regions. Buried among the debris on his chronically disorganized desk was a paper that Bob Duce, an atmospheric chemist now at Texas A&M University, had recently sent him. In it Duce estimated that around half the iron in the open ocean comes from dust settling out of the atmosphere, as opposed to being upwelled from the deep with the other nutrients. When Martin looked again at Gordon's iron measurements, he realized that, if anything, Duce was underestimating the contribution of atmospheric dust. Far from shore, there was so little iron in deep water that upwelling currents could not bring up much. Around 95 percent of the iron at the surface there must come from dust falling out of the sky.

With the distribution of iron thus controlled by the geometry of winds and land masses, particularly arid land masses, it became possible to imagine that some regions of the ocean might be drastically shortchanged. Others, meanwhile, would be favored with copious rains of iron. In the equatorial Atlantic, for instance, winds sweeping off the Sahara carry dust as far west as Barbados. All along that path, there is no shortage of iron in the water, and, Martin noted with interest, no shortage of phytoplankton. The equatorial Pacific, on the other hand, was known even before Martin got interested in the subject as a region of exceptionally low dust fallout. It is also the second largest region of high nutrients/low chlorophyll. The largest such region is the Southern Ocean, and the winds blowing over it cross little land and even less that is not covered by ice. Thus they drop almost no dust and almost no iron on the water.

Just reading the literature, then, was enough to give Martin a nice, simple hypothesis: no dust means no iron and no plankton, or less than there should be. But the idea still had to be tested in the field. Between 1987 and 1990 Gordon and Fitzwater did just that, in the Gulf of Alaska, the equatorial Pacific, and the Antarctic. On all those cruises they brought their own clean minilab with them—all plastic, positive pressure, the size of a small house trailer—and set it up on the deck of the ship. They used special 30-liter sampling bottles that went through the sea surface closed (so their insides would not get coated with bilge water or fuel) and then snapped open automatically at depth. They lowered each bottle on Kevlar line, because ordinary steel wire might have contaminated the water around the bottle. And when the bottle came back up, now weighing more than 100 pounds, Gordon and Fitzwater either loaded it onto a plastic cart or manhandled it directly over to the clean lab—whatever they did, they did not let it touch the deck, because it might pick up a few atoms of iron that it would take back down on the next trip.

With those clean samples Martin's colleagues did a simple experiment, again and again. First they measured the amounts of iron and nitrate in the water. In all three high-nutrient/low-chlorophyll regions they found that the plankton did not seem to have enough iron to bloom. Next Gordon and Fitzwater fertilized some of the samples with a minute amount of iron, while leaving the others alone. Almost invariably they got the same results: In the unfertilized bottles, not much happened. In the fertilized bottles, though, the iron and nitrate concentrations went down, sharply, and the chlorophyll concentration went up, until after six days it was many times higher than in the control bottles. The fertilized water turned greenish, and when you held the bottle up, you could see light glinting off innumerable floating cells. The iron supplement had caused the phytoplankton to bloom.

The most striking results came from the Southern Ocean. Every year, during the six months of sunlight there, phytoplankton bloom profusely in a narrow band near the Antarctic coast, along the edge of the sea ice. Those blooms support vast schools of krill, which in turn feed seals, penguins, and whales. When Gordon and Fitzwater added iron to a bottle of nearshore water in the Ross Sea, the iron did not do much—

because there was already plenty in the water, much of it from bottom mud. But a few hundred miles offshore, the story was different. Iron added to a seawater sample there triggered a fourfold increase in the amount of plant matter in the bottle.

The open Southern Ocean is 8 million square miles of nutrient-rich water, fed by the most vigorous upwelling currents in the world, with temperatures no colder and sunlight no fainter than near the coast, but for lack of iron the phytoplankton do not bloom. Yet as Martin soon found out, that was not always true in the past. Not long before Martin had gotten interested in the subject, French and Russian researchers had extracted a two-kilometer deep core from the Antarctic ice—a frozen record of Earth's atmosphere through the last ice age and beyond. That core showed, as had previous ice cores, that the carbon dioxide level in the atmosphere had been much lower at the peak of the last glaciation, 18,000 years ago: only 200 parts per million, compared with 280 parts per million in the preindustrial nineteenth century and 360 parts per million today. At the same time, though, the amount of dust falling onto Antarctica was between 10 and 20 times greater. The world was drier during the ice age, with vast tropical deserts and stronger winds that carried dust off the deserts and out over the sea. Along the whole length of the ice core, Martin noticed, atmospheric carbon dioxide and atmospheric dust fluctuated in counterpoint; one went up when the other went down.

For Martin the explanation was simple: whenever a lot of iron-rich dust had fallen out of the atmosphere, the Southern Ocean phytoplankton had been cured of their anemia, and the massive blooms that ensued had drawn carbon dioxide out of the atmosphere. That is what made the world colder during the ice age: the phytoplankton did it. By blooming to their full potential and converting all the available nitrate into organic matter, they drew two billion tons of carbon dioxide out of the surface water, which drew it out of the atmosphere. Martin calculated that it would not have taken a lot of iron to get the phytoplankton going; around 300,000 tons scattered over the Southern Ocean over the course of the growing season would have been enough. And that scale of iron fertilization, he pointed out, does not require geologic forces. We could do it ourselves—and if we were serious about combating the greenhouse effect, we ought to consider it. "You give me half a tanker full of iron," Martin told a *Washington Post* reporter, "and I'll give you another ice age."

He did not really believe anything so extreme. Martin knew that ice ages were more complicated than that; what triggers them is still not understood completely. Earth's orbital cycles play a role, and so do ocean currents (see chapter 10). Martin must also have known that his remark to the *Washington Post* would make the average biological oceanographer's hair stand on end. The average biological oceanographer would not say plankton are "cruddy things," even in jest, and does not believe they or the sea should be tampered with lightly. Even from a purely intellectual standpoint—leaving aside Martin's suggestion for a greenhouse fix—the iron hypothesis was bound to be unappealing to many of his colleagues. It was simple to the point of sounding simplistic. When you have spent years or decades studying some small aspect of the sea and finding it so complicated that even the most basic fact about it eludes you—well then, you are not likely to respond favorably when a cranky graybeard rolls up to you in his wheelchair, fixes you with a malicious twinkle and says, in his gravelly voice, "Aw baloney, it's just iron deficiency."

But Martin was like that. He liked simple arguments rather than complex ones, he had a strong streak of iconoclasm, and he had an outsider's temperament—he did not seem to feel he belonged to the club of which he was in fact a distinguished member. When that club convened at a San Diego hotel in early 1991, under the auspices of the American Society of

Limnology and Oceanography, for a special symposium on Martin's hypothesis, Martin sat in the back of the room with Gordon and Fitzwater. Fidgeting impatiently, he occasionally took potshots at the speakers. Some of them, in turn, could barely conceal their outrage at his idea, and especially at the attention it was getting; by then Martin had appeared on *Good Morning America*. But at some level he did not mind being in a room full of people who thought he was full of crap. His role model, after all, had been Clair Patterson. And after several days of wrangling, Martin came away with what he wanted: his peers' endorsement of an experiment that would test his hypothesis once and for all. That was his real goal, he had once said—not to save the world from global warming, but simply to prove that phytoplankton are indeed iron-deficient. After that he would retire, perhaps to a quiet life of collecting copepods. Or so he said.

The main criticism of Martin's previous experiments had been that his bottles were too small to reflect the reality of the ocean. A 30-liter bottle, it was said, would generally fail to capture the zooplankton that graze on large phytoplankton such as diatoms, which were just the ones that grew most luxuriantly in Martin's iron-spiked bottles. If you fertilize a pasture with iron but then lock out the sheep, critics argued, the grass will grow taller, whether the iron had helped or not. The only way to refute that criticism definitively, Martin realized, was to do the experiment in the ocean itself. He was organizing that experiment—the first controlled manipulation of a patch of open ocean—when he died in 1993.

Keeping track of a turbulent patch of seawater for a week, long enough to see how the plankton might react to a minute iron supplement, was not going to be easy. Martin's plan was to tag the patch with a harmless chemical, sulfur hexafluoride, that is much easier to detect than iron. It would act as a kind of invisible dye. For 24 hours the research ship would steam back and forth across a five-mile-wide square, taking in seawater, mixing it in a tank with iron sulfate, mixing that in another tank with sulfur hexafluoride, then flushing the lot out into the ship's wake—half a ton of iron in all, spread so fine that its concentration in the water would never be more than a few parts per billion. After that the ship would follow the drifting patch, to see how the plants responded.

The first time this was tried, in the equatorial Pacific 300 miles south of the Galápagos, the plants responded immediately but modestly. The concentration of chlorophyll inside the patch nearly tripled. The water turned a bit greener. The plants drew a bit of carbon dioxide out of the atmosphere but not much. Apparently the heavy iron, added all at once at the beginning of the experiment, sank out of the surface water before the phytoplankton there could use it all up; and on the sixth day, unluckily, a front of light, low-salinity water moved in and dunked the iron-enriched patch out of the sunlight. That alone would have nipped any plankton bloom in the bud.

So in 1995 Martin's successors at Moss Landing, Kenneth Coale and Kenneth Johnson, along with Fitzwater, Gordon, and a shipload of other oceanographers, went back to the equatorial Pacific to try for a more conclusive test of the iron hypothesis. This time they spread their half-ton of iron in three doses over the course of a week. And this time the ocean cooperated. The iron-enriched patch remained largely intact, even as it drifted 950 miles to the southwest. The patch was like a giant bottle—and the results of the fertilization were like the results of Martin's bottle experiments: the phytoplankton population exploded. With the extra iron, each cell immediately made more chlorophyll, which allowed it to capture more sunlight. The growth rate of the phytoplankton in the patch doubled, to nearly two cell divisions per day. The chlorophyll concentration shot up by a factor of 27. You did not not need fancy measurement equipment or a Ph.D. in biological oceanography to detect that kind of change. "It was almost biblical," Johnson told the British magazine *New Scientist*. "In a week the ocean went from a desert to a jungle, from clear blue to dark green."

Even on board the ship, though, there was ambivalence about that success; fear that this newfound skill at manipulating the ocean might reinforce the tempation to try it on a massive scale, with half a tanker of iron rather than half a ton. "None of us was really prepared for what it would look or feel like," Coale said to *Science News*. "There were some of us who were quite pleased and others of us who would walk out on the fantail and burst into tears. It was a profoundly disturbing experience for me. We had deckhands come up to us and ask, 'Did we do this?' "

Explaining why the ocean was a desert in some places and a jungle in others was what Martin's hypothesis had been all about, and the explanation had turned out to be as simple as he said it was. It was after all just iron deficiency; just adding iron had made the desert bloom. But the bloom was selective. Before the experiment the water had been dominated by the small cyanobacteria, *Prochlorococcus* and *Synechococcus*, which were apparently already growing as fast as they could. The diatoms must have been more desperate for iron, because as in the bottle experiments, it was they that really took off: their biomass increased by a factor of 85. The copepods that would normally crop them back could not reproduce that fast. It seems, then, that the amount of iron in the water helps determine not only how much plankton is present but also what kind.

Some kinds of phytoplankton emit a compound called dimethyl sulfide (DMS), and they were among the ones stimulated by the iron; the concentration of DMS more than tripled in the fertilized patch. DMS is a breakdown product of a compound that helps phytoplankton regulate their internal pressure and keep sea salt from invading and poisoning them. It also seems to be part of a chemical defense that poisons grazing zooplankton. But once it gets into the atmosphere, DMS has an entirely different effect. There it is converted to sulfate particles. They absorb sunlight, and they also seed the formation of clouds—water droplets condense around the sulfate particles—which reflect sunlight back to space. Both processes, in theory, help to cool the planet—to counteract the greenhouse effect with a "parasol effect." If iron fertilization could strengthen Earth's parasol with more DMS, it would represent an added bonus that Martin did not even anticipate.

But he seems to have been right about the main thing: dumping iron could weaken the greenhouse. The massive bloom triggered in the second fertilization experiment drew about 3,000 tons of carbon out of the surface water. Coale and his colleagues did not attempt to measure how much of that fixed carbon sank out of contact with the atmosphere. But much of it was in the form of diatoms, which tend to clump together and sink; fresh green gobs of diatoms have been seen on the seafloor after natural blooms. In the Antarctic, sinking currents of cold water would help carry any carbon fixed by plankton into the abyss. The Moss Landing researchers hope to do a fertilization experiment there in 1999 or 2000, as the ultimate test of Martin's theory. The most plausible hypothesis at the moment is that fertilizing the Southern Ocean in a massive way would indeed draw a substantial amount of carbon dioxide out of the atmosphere.

Enough to bring on an ice age, or at least to alleviate global warming significantly? That is not clear. Sediment cores extracted from the South Atlantic show traces of massive plankton blooms at the height of the last ice age—and, simultaneously, massive infusions of iron from deserts in Patagonia. But the blooms do not seem to have covered the entire Southern Ocean, and so it is not certain they would have been enough to lower the atmospheric carbon dioxide concentration from 280 parts per million to its ice age level of 200 parts per million.

As for the possible effects of man-made iron fertilization, Jorge Sarmiento, a Princeton University oceanographer, tried to simulate them, for the 1991 San Diego meeting, with a computer model of the ocean and atmosphere. He found that iron fertilization of the entire Southern Ocean every year for the next century—the iron keeps getting used up, so you have to keep adding it—could remove around 17 percent of the carbon dioxide our smokestacks and tailpipes would be belching out under the "business-as-usual" scenario. (In that scenario, CO_2 emissions continue to grow at the rate they were growing in 1990.) Instead of around 785 parts per million in the year 2090, the atmospheric concentration would be around 715 ppm, or twice its current level. If on the other hand we managed to freeze our carbon dioxide emissions at their 1990 levels, then fertilization might keep the atmospheric concentration to around 440 ppm instead of 500 ppm.

To Sarmiento and to many other oceanographers at the San Diego meeting, this sounded like a relatively small benefit, not worth the risks. In 1997 a new idea emerged that suggests a way to increase the benefit: scatter iron not just over the Southern Ocean but over the whole world ocean, or most of it. The author of the theory is Paul Falkowski of the Brookhaven National Laboratory, who had been one of the participants in the triumphant iron-fertilization experiment. Falkowski argues that it is not just the high-nutrient/low-chlorophyll regions that are iron limited. The central basins of all the oceans, far from land and from any upwelling currents, are also low in chlorophyll, simply because they are low in the conventional nutrients, nitrate and phosphate. It is nitrate that is the real limiting factor in those vast regions, Falkowski thinks, and it in turn is limited by the amount of iron that is available.

The main source of new nitrate in the open ocean, Falkowski points out, is a genus of cyanobacteria, *Trichodesmium*, that has the rare ability to "fix" pure nitrogen from the atmosphere. Nitrogen makes up 78 percent of the atmosphere. To make the nitrogen-fixing enzyme, though, *Trichodesmium* needs a lot of iron. Thus if there were more iron in the middle of the ocean, there would be more *Trichodesmium*, more nitrate, and more other types of plankton. By Falkowski's calculations, fertilizing the central ocean basins as well as the high-nutrient/low-chlorophyll regions with iron—if so huge a project were practical, which it probably is not—could in principle remove more carbon dioxide from the atmosphere than we have put into it since the start of the Industrial Revolution. Something approaching that, he says, may have happened naturally in the last ice age, through the action of wind-blown dust.

Falkowski is not recommending massive iron fertilization; he says he has no opinion on the subject. Most biological oceanographers shudder at the thought of treating the ocean like one giant farm. But the question of whether to fertilize will not stay in their hands for long—which is precisely their worst fear. In the wake of the successful iron experiment, commercial operators in the United States and Chile have started their own tests: they are hoping that iron fertilization will produce more fish, with a weaker greenhouse effect as a fringe benefit. And the Japanese electric power industry, a leader in the effort to find ways to trap carbon at the smokestack and put it at the bottom of the ocean, has launched an iron-fertilization effort. Given how hard it is to limit pollution, even the small reduction in atmospheric carbon dioxide predicted for fertilizing the Antarctic does not seem uninterestingly small to some people.

Is it wise, is it moral even, for us to attempt to fix nature on so grand a scale? To tinker deliberately with the biogeochemical cycles and the climate of the entire planet? John Martin himself was nervous about the idea. But on the whole he thought it preferable to the unintentional and careless tinkering we are doing already. From his lab at Moss Landing, he used to look up at the twin stacks of the Pacific Gas and Electric plant, which rise like cathedral spires above the little fishing village on Monterey Bay. During the summer, when air conditioners in central California are pumping furiously, the PG&E stacks pump more than 4,000 tons of carbon dioxide a day into the atmosphere. Martin was skeptical about the prospects of our controlling our appetite for fossil fuels anytime soon. Even after the landmark Kyoto agreement of December 1997, in which the industrial nations agreed to turn the corner at last—to reduce their emissions of greenhouse gases, by the year 2012, to five percent less than they were in 1990—it is still not clear that Martin was wrong. The Kyoto agreement must be ratified by legislatures, and it does not constrain developing countries such as China, which by the middle of the next century will probably have passed the United States to become the world's biggest polluter. Martin believed that experimental fix-it schemes like iron fertilization, crazy as they might sound, might one day be necessary.

"We're already involved in the biggest experiment ever," he said in 1990. "We're finding out what's going to happen if we dump three billion tons of CO_2 in the atmosphere every year. That is the biggest manipulation of the environment ever. What I argue is we're going to keep doing this experiment. And we'd better know about ways, if we have to, to bring CO_2 out of the atmosphere.

"The Chinese are not burning their coal right now—they have massive coal reserves. As they industrialize they're going to want to use their coal. They're not going to say 'Gee gang, we're going to use solar power so we don't put out greenhouse gases.' The average Chinese peasant wants to have electricity, and a refrigerator, and maybe a motorbike—these people want to have the same crap we've got, and they're going to use what they've got available. CO_2 is not going to go down next year, or level off next year. It's going to go up by at least 3 billion more tons, and the next year 3 billion more and probably 5. So by 60 years from now, you may be going up 10 or 15 or 20 billion tons a year.

"The purists say iron fertilization is bullshit, and we shouldn't even think about it. 'This is horrible,' they say. 'We should be conserving.' That's very true. But what if we get into a situation where indeed global warming does occur? And all of a sudden you start melting back the permafrost, and you release even more and more CO_2 and methane, because you've got all this organic matter stored in the permafrost. You may suddenly say 'We've got to pull CO_2 out of the atmosphere.' Then if you know something about iron, you can say 'We know how to do this.'

"I'd see an ethical problem if we can go down and fertilize the Antarctic and remove the CO_2, but in the process we're going to kill all the whales and the penguins. Then I'd say 'Don't do it. Man has created this, he can stew in his own juice. Leave the whales alone.' But on the other hand if it turned out that you could go down there and add the iron and not disturb the environment, then I'd say go ahead and do it."

Adding shiploads of iron to the ocean will inevitably disturb it somehow, of course; and unfortunately, we will not really know how until we do it. Martin liked to joke that iron fertilization might actually produce a baby boom of whales in the Southern Ocean. If the iron triggered diatom blooms, as it did in the equatorial Pacific, that could indeed lead to more krill, which eat diatoms, and thus to more seals, penguins, and whales. But perhaps in the Southern Ocean something other than diatoms would respond best to the iron, something krill do not eat; or perhaps the diatoms would bloom but the krill would be outcompeted for them by salps. Perhaps, in the end, the existing food web would collapse. The idea that phytoplankton are limited by a lack of iron is a simple truth, but the marine food web is too complex for oceanographers to predict in advance how an iron supplement would change it. In Sarmiento's simulation, iron fertilization caused a large part of the southern Indian Ocean to become anoxic, that is, depleted in oxygen. Such a condition would not be congenial to the animals that live there.

We have some experience with that sort of thing already, because we are fertilizing the ocean already on a large scale—not with iron but with nitrogen and phosphorus. All over the planet, sewage from our homes and fertilizer from our fields and effluents from our factories flow into coastal waters, at a rate that is steadily increasing as the human population does. The nutrients stimulate phytoplankton blooms. In the Gulf of Mexico, which thanks to the Mississippi River receives runoff from two-fifths of the contiguous United States, some researchers think the added nitrogen has increased fish stocks. But ever since the Mississippi floods of 1993, the gulf has been visited by a more worrisome phenomenon: a zone of oxygen-depleted water the size of New Jersey, near the seafloor around the mouth of the river. The "dead zone" has returned each summer, and the timing of it seems linked to river-borne pulses of nitrogen. All the plant matter engendered by that nitrogen gets eaten by animals and transformed into feces, or else it falls uneaten to the seafloor; either way, the organic matter is degraded there by bacteria, and that process draws oxygen out of the water. So far, the only things affected by the dead zone are seafloor creatures that cannot swim away, such as polychaetes and other worms. Not many people care about seafloor worms.

Some plankton blooms do affect people directly, however. Those are the ones that are commonly known as red tides, even though they have nothing to do with tides and only a few are red. Harmful algal blooms, as scientists call them, are natural. They have been with us since Exodus, when Moses, by way of showing God's power to Pharaoh, "lifted up the rod, and smote the waters that were in the river . . . and all the waters that were in the river were turned to blood. / And the fish that was in the river died; and the river stank. . . ." Red tides have even been with us since well before biblical times, according to the fossil record. But in recent decades there has been a dramatic increase in their frequency. It may be in part that we are more aware of them and are watching them more closely, now that we understand how the toxins released by some red-tide organisms can pass through shellfish and fish to man. Many scientists believe, however, that the proliferation of harmful blooms must have something to do with our intensive fertilization of coastal waters. We are creating a different environment in which different species will thrive, they say—and sometimes the change will not be a happy one.

Certainly nutrient pollution is the leading hypothesis to explain one of the more spectacular red tides of recent years, the outbreak of *Pfiesteria piscicida* off the coast of North Carolina and in the Chesapeake Bay. *Pfiesteria* is not a new organism, anymore than other red-tide plankton are, but if it had bloomed in the past the way it has in the late 1980s and 1990s, we would not likely have missed it. Although *Pfiesteria* is a dinoflagellate, it is not one of the photosynthesizing kind; it is more animal than plant, in that it makes a living by eating other organisms, such as bacteria and phytoplankton. Many dinoflagellates do that, but *Pfiesteria* is unusual: at certain stages of its life cycle, it also preys on fish that are 10,000 times its size. Its life history is more like a web in its complexity than a simple cycle. JoAnn Burkholder of North Carolina State University, who discovered *Pfiesteria* in 1988, has identified at least 24 distinct stages, and the organism seems to move back and forth among these stages depending on the conditions in its environment.

Its favorite environment is shallow, slow-moving, brackish

estuaries. There it lies waiting in the bottom mud in dormant cysts—waiting for fish to swim by. Something excreted or secreted by the the fish awakens it and causes it to transmute into a more conventional-looking dinoflagellate. Propelled by its long flagella, *Pfiesteria* then swims toward its prey. As it does, it releases into the water a toxin potent enough to stun the fish and keep them from fleeing; potent enough to open bleeding sores in their skin and to cause the skin to slough off; potent enough, finally, to kill the fish outright. In laboratory experiments Burkholder has found no fish that survives exposure to *Pfiesteria*; it kills everything from blue crabs to largemouth bass. After she and her colleagues discovered that the dinoflagellate's toxin can become airborne—they discovered that by contracting a range of fairly serious health effects, from asthma to nausea to memory loss—they started working with gas masks and converted the laboratory into a "biohazard level III" facility, as if they were working with a lethal virus. In its natural habitat, once *Pfiesteria* has incapacitated a fish, it attaches to bits of sloughed-off skin and feeds on them. It also has the ability to transform into something like an amoeba, the better to crawl over the corpse of the fish.

Pfiesteria had already been linked to extensive fish kills in North Carolina when it broke out dramatically in Chesapeake Bay in the summer of 1997. With reports coming in not only of dead fish but also of sick fishermen and swimmers, state officials closed sections of three rivers that empty into the bay. It is tempting to draw a moral from all this—to see the attack of so savage and alien a beast as a kind of retribution for our sins. Burkholder herself is convinced that pollution, especially runoff from hog and chicken farms, is responsible for the sudden *Pfiesteria* plague. Pollution helps *Pfiesteria* indirectly, she says—the inorganic nutrients stimulate blooms of phytoplankton, which it eats when fish are scarce—but also directly: Burkholder has found that the dinoflagellate is especially abundant around sewage outfalls. She believes it can feed on the organic matter coming out of the pipe. So far, though, she has not persuaded all her peers that the *Pfiesteria* problem is a human creation, or even that it is responsible for all the fish kills she has blamed it for.

Most scientists would agree, however, that reducing coastal pollution would be a sound idea. Yet it is not one that is about to be realized. As a result, at least one leading expert, Donald Anderson of Woods Hole, argues that we should be attacking red tides as we would agricultural pests on land—in effect counteracting our pollution of coastal waters by manipulating them in a more planned way. We might deploy viruses that target harmful phytoplankton, or parasites, or even chemical herbicides if they could be made specific enough. In 1996 off the coast of Korea, according to Anderson, 60,000 tons of clay were dumped into 100 square kilometers of ocean to control a red tide that was threatening fish farms; the idea was that the clay would strip plankton from the water as it settled harmlessly to the bottom. "I believe that some harmful algal blooms can be controlled or managed," Anderson wrote in a 1997 article in *Nature*,

> . . . economically and without disastrous environmental consequences. This belief may brand me as a heretic among my colleagues, some of whom fear that the ocean will be further despoiled by inept human attempts to manipulate ecosytems we do not understand.
>
> At the heart of this negativism is a conviction that mankind does not possess the skills, knowledge or right to manipulate the marine environment on any significant scale. We are, however, already doing exactly that.

Anderson's argument is that of John Martin: it makes no sense, given how much damage we are doing to the ocean, to shrink from trying to fix it. And although the argument may seem heretical in the small community of biological oceanographers, it is hardly out of step with the times. It is probably inevitable that we will attempt to control and manage the ocean, or parts of it anyway, as we now do the land, with such mixed results. But is it negativism or is it realism to think that in the ocean, where our ignorance is more extreme, the technological fix is even more likely to give way to the law of unintended consequences?

One of the weirder demonstrations of that law comes, as it happens, from the Antarctic, the most likely site for iron fertilization. In the 1930s, when chlorofluorocarbons were invented, and in the ensuing decades, when they became known as miracle chemicals that could safely be used in everything from refrigerators and air conditioners to foam cushions to deodorant cans, who could have guessed that one day, thanks to them, we might be doing serious damage to phytoplankton in the Southern Ocean? And yet the fact that chlorofluorocarbons are responsible for the hole in the ozone layer over Antarctica has now been established; and also the fact that ultraviolet light pouring through the ozone hole can inhibit photosynthesis, and damage plankton in other ways as well. One study published in 1992 found that phytoplankton in the fertile zone along the edge of the sea ice were already producing between 6 and 12 percent less plant matter. What effect that might be having on the rest of the food web, from krill on up to whales, remains unknown.

There is other evidence, as we contemplate fertilizing the Southern Ocean, that plant productivity there may recently have decreased, and for reasons that have nothing to do with the ozone hole. William de la Mare, a researcher at the Australian Department of the Environment, reported in 1997 that the ring of sea ice around Antartica had retreated abruptly, between the mid-1950s and the 1970s, by nearly 3 degrees of latitude; its summertime edge had moved southward from 61.5

degrees south, on average, to 64.3 degrees. As a result, according to de la Mare, sea ice now covers 2.2 million square miles less of the Southern Ocean—a 25 percent decrease—and the fertile zone for phytoplankton is correspondingly smaller. One might assume that the decrease was caused by man-made global warming, but that is not clear. It may have been a natural fluctuation, albeit a surprisingly abrupt one.

The change had not been noticed before because the first satellite capable of monitoring sea ice was not launched until 1973. De la Mare detected it by an ingenious method that was almost as interesting as the result itself. Before satellites came along, it turns out, the best records of Antarctic sea ice were kept by whaling ships. Whaling began in Antarctica in 1904, and the first factory ships arrived in 1905. The whalers quickly discovered that whales, like phytoplankton and krill, tend to be concentrated around the ice edge. As they recorded the position of each catch, they were therefore also recording the extent of the ice. De la Mare went to the trouble of retrieving those records from the International Whaling Commission in Norway and extracting from them the southernmost catches for each year and each longitudinal sector. The data revealed that, between the 1950s and the 1970s, the whalers had moved progressively farther south as they followed the retreating ice edge.

Those climatologically valuable records are about all that is left of Antarctic whaling. By the late 1950s the whalers had nearly wiped out the stocks of blue, humpback, and fin whales in the Southern Ocean. They then turned their attention to sei and finally to minke whales, until commercial whaling was banned altogether in 1987. Now the most important fishery in the Southern Ocean is for krill, the things the whales used to eat. All the unintentional damage we have inflicted on the ocean through pollution, and may yet inflict as we begin to "manage" and "control" it, is nothing compared to these deliberate effects of whaling and fishing. It is because of them above all that the sea around us at the end of this century, though it may look as wild and untouched as ever, is vastly different from the one we knew at the end of the last.

CHAPTER 8

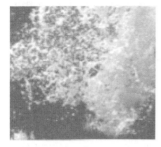

Twilight of
the Cod

In the Massachusetts State House, high above the gallery in the House of Representatives, directly opposite the painting of John Hancock proposing the Bill of Rights, there hangs a wooden, five-foot-long codfish. It is painted gold with scarlet gills, and it has been there for a century—ever since it was moved from the old House chamber, where it had hung for a century before that. The transfer of the Sacred Cod on March 7, 1895, was an occasion for pomp and soaring oratory. A committee of 15 legislators was appointed to fetch the fish. Two by two they followed the sergeant-at-arms into the old chamber, watched as the cod was lowered onto a bier draped with the American flag, and then marched behind the four pages who carried it into the new hall. There the cod and its entourage were greeted with a deep bow by the senator from Gloucester, the state's preeminent fishing port. The rest of the assembly rose to their feet and applauded the fish vigorously. "Everybody who could make a pretext for touching its fins or for holding it straight on the stretcher did so," the *Boston Daily Globe* reported the next day. The triumph of the codfish was front-page news in both the *Globe* and the *Boston Herald*; each devoted nearly half a broadsheet to the event.

The early 1890s were good years for cod fishing in Massachusetts, and in particular for Gloucester. At the 1893 World Columbian Exposition in Chicago, where America was introduced to electricity, Gloucester mounted an elaborate exhibit, featuring a scale model of its thriving waterfront. That same year the Portuguese immigrants to that waterfront finished building themselves a church, Our Lady of Good Voyage, and topped it with a gaily painted statue: Madonna with schooner. Rudyard Kipling was holed up in Brattleboro, Vermont, in the 1890s, writing *Captains Courageous*, his paean to the Gloucestermen who went down to the sea in schooners and dories. Sailing on the rich offshore banks, from Georges Bank off Cape Cod to the Grand Bank off Newfoundland; staying at sea for months; fishing with hook and line from small boats tossed on large waves, those men sustained an industry whose reach was global. In 1895, fishermen caught 60,000 tons of cod in the waters off New England. In May of that year, two months after the shifting of the Sacred Cod, one man landed the Patriarch Cod—a six-foot-long, 211-pound fish. Its like has never been seen again.

Fishing has changed a lot in the last century. On a bleak sleety morning in late 1994, a few dozen descendants of the captains courageous gathered 15 miles inland from Gloucester, and about as far from Kipling's fish pier as it is possible to get. They came, in their flannel shirts and jeans and baseball caps, to a Holiday Inn set in the strip-mall ugliness of Route 1 in Peabody. They sat, in a pink, drop-ceilinged ballroom, under a reflecting disco ball, and listened to their fate being discussed by men in suits—the Groundfish Committee of the New England Fishery Management Council. They watched, more or less mutely, a computer-model presentation of the options open to this committee. The presentation was scarcely comprehensible even to scientists in the audience, and the discussion of the committee was lackluster and at times non-existent. But it mattered little: Everyone knew that the options were all but nonexistent too. A month earlier, the council had decided that fishing for cod on Georges Bank—as well as for haddock and yellowtail flounder, the two other important bottom-dwelling fish—must essentially be stopped. The committee's task was to work out the details.

All over the world similar committees are facing similar Hobson's choices. At the turn of the century the world's fishermen pulled around 5 million tons of fish out of the ocean; by the late 1980s the total was 86 million tons. But since then, with no decrease in fishing effort, the catch has dropped to around 80 million tons—a sign that limits have been reached. Actually they were reached much earlier for many of the fish we are most familiar with; for years the growth in catch had been sustained by aggressive exploitation of small fish, like pilchards and anchovies, that are mostly ground up for animal feed. The peak catch of haddock, a close relative of cod and a longtime rival for dinner-plate hegemony in New England, occurred in the 1920s. The heyday of halibut, a flat, bottom-dwelling monster that used to grow to 400 pounds in Massachusetts Bay, was in the mid–nineteenth century. No one fishes for halibut off New England anymore; that stock is all fished out.

And now cod—never a delicacy, ever a staple among fish— are close to the same wretched end. On that happy afternoon a century ago in the State House it could not have seemed possible. "This sedate and solitary fish," Representative James Gallivan of Boston had told the assembly, speaking of the wooden one, ". . . commemorates Democracy. It celebrates the rise of free institutions. It emphasizes progress. It epitomizes Massachusetts." This was not just posturing. Endless resources, free for the taking, are what defined America, and it started with cod. Cod helped spur the settlement of the New World, from Newfoundland down to New England. They were its first industry and export. They fed the Pilgrims. But now, after 500 years, from Georges Bank right up to the Grand, the cod are all but gone.

Atlantic cod, *Gadus morhua*, have been around a lot longer than we have, probably more than ten million years. Pacific cod originated more recently, between three and four million years ago, when continental drift separated Asia from North America; at that point a small party of Atlantic pioneers swam through the Bering Strait into the Pacific and, as it turned out, founded a new species. Not long afterward Earth was plunged into the Pleistocene ice ages, and the Arctic Ocean became too cold for Atlantic cod. Isolated from one another by ice, the Atlantic and Pacific fish stopped interbreeding and evolved apart.

Cod survived even the ice ages, though, presumably by moving south. In the Atlantic they live today from the Barents Sea north of Norway down the European coast as far south as the Bay of Biscay and from northern Labrador and Greenland down the North American coast as far as Cape Hatteras. As far as biologists can tell, the cod that live on opposite sides of the Atlantic and even at different points along the North American coast form distinct stocks, or populations. But they are still in occasional touch with one another and still belong to the same species. In 1961, for instance, a fish that had been tagged by British researchers in the North Sea four years earlier was caught off Newfoundland, after a journey of more than 2,000 miles.

Cod live in coastal waters, a thousand feet deep or less, because that is where they find food: the web of small animals that is supported by phytoplankton. In shallow water, tides and currents can readily stir nutrients from the bottom mud up to the sunlit surface where phytoplankton bloom. The chain of shallow offshore banks that runs along the coast from New England to Newfoundland is especially plankton-rich; since the last ice age, it has been an archipelago of productivity, and in particular of cod.

The banks themselves were created in some earlier glaciation; when is not clear. At that distant sometime, between a hundred thousand and perhaps five million years ago, the sea level was much lower than it is today, because the water was stacked up in continental ice sheets. The continental shelf that is now submerged was then a flat coastal plain dipping gently toward the sea. Many rivers meandered over this plain, and over the millennia their wandering channels dug a series of basins into it. Along the ancient coast, the rivers left a chain of low hills, steep toward the land and gently dipping toward the sea, and punctured by just a few large estuaries. In the last ice age, glaciers surged right up to the edge of the hills and dug the basins even deeper. When the ice sheets finally melted and the sea level rose, the basins were flooded and the hills became submerged offshore banks. The Gulf of Maine is one such basin; Georges Bank, at the mouth of the Gulf of Maine, is one such bank. The banks are often only a few tens of feet deep at their crests and never more than a few hundred. They are covered with layers of sand and gravel dumped by the last ice sheets.

Cod are groundfish—bottom dwellers—and have the mottled coloring of sand or gravel. In other ways too they are not the stuff of poetry. They are large, typically two to three feet long and eight pounds or so at maturity, but not exceptionally large; strong swimmers, with their powerful tail fin, but not exceptional ones; agile swimmers, thanks to fins on the back and belly that act as rudders, and ones on the sides that act as horizontal thrusters—but again, not exceptional. When you look at a cod in a tank and it looks back at you with its big round eyes, you think: Fish. Cod are a kind of essence of fish, a default setting from which other fish are extravagant variations. Cod are generalists. And they are survival machines.

For one thing, they are omnivorous. Adult cod favor small schooling fish—capelin off Newfoundland, herring on Georges Bank—but they also eat crabs, shrimp, squid, and clams. (They swallow six-inch clams whole and digest out the meat;

half a dozen shells may be found in a cod's stomach, stacked like saucers.) For another, they live a long time, 20 to 25 years if left to their own devices, and grow to adulthood fairly quickly—around age three on Georges Bank, and around seven in the colder waters off Newfoundland.

From that point on the fish spawn copiously. Around spawning time, a large fraction of a female cod's weight is eggs. She releases as many as several million at a throw, and her partner more than matches that output in sperm. Some decades ago a biologist in Newfoundland named Vivien Brawn observed cod spawning in a laboratory tank (which is not easy; the fish require almost total darkness to perform). It is surprisingly touching, as she describes it—not anonymous at all:

> The male fish, on seeing the female, raised his dorsal fins momentarily, or paused, and then approached the female slowly. Positioning himself in front of the female and about a foot away, the male cod began the courtship flaunting display. . . . All the median fins were fully erected and the male made many exaggerated lateral bends of the body. . . . Swimming with an excited, jerky, undulating movement with many unnecessary circles, the male cod began to move away from the female. . . . The female swam slowly in the male's general direction but did not attempt to follow his elaborate path. . . . If [she] swam toward the edge of the territory she was pursued and overtaken. . . . The flaunting of the male was accompanied by a low grunting sound. . . . At each grunt the female showed an increase in excitement. . . .
>
> Sometimes the female followed the male more and more sluggishly until she became stationary near the bottom. The male . . . then swam under the female prodding [her] with his snout in the ventral region and giving a loud grunt. . . . A ripe female invariably swam rapidly up to the surface . . . closely followed by the male and both fish made many vertical circles to the surface and down again. . . .
>
> Eventually the female came to rest at the top of the tank with her back just breaking the surface. The male approaching the female from behind or one side, pressed the lower jaw onto the

back of the female and, pushing her downwards, swam onto her back, grasping the female with the pelvic fins just posterior to her head. . . .

Once the male had mounted dorsally, he immediately slipped down one side of the female, still with his ventral surface closely pressed against her body and clasping her with the pelvic fins. The male came to lie in an inverted position below the female with the ventral surfaces of both fish and their genital apertures closely pressed together.

On being mounted the female stiffened her body and swimming with awkward sideways sweeps of the tail, contracted the body ventrally and almost immediately spawned. The male also swam and spawned during the ventral mount. The combined movements of the tails drove the pair round in a horizontal circle at the surface and doubtless served to mix the eggs and sperm.

The whole process takes about 15 minutes.

It has been observed in the wild as well, although not nearly on so intimate a scale. Spawning cod come together near the seafloor in huge schools that show up clearly on sonar records—or rather they used to when there were a lot of cod. George Rose of the Canadian Department of Fisheries and Oceans (DFO) in St. John's, Newfoundland, has seen such schools, hundreds of millions of fish spaced about a body length apart, at a depth of a thousand feet just north of the Grand Banks. The last one he saw, in 1992, was a mile and a half across and was shaped like an upside-down saucer, domed at the center. For the first ten days after Rose discovered it, the school stayed more or less put near the edge of the continental shelf. But every now and then thin, transient columns would sprout up toward the surface from the main mass of fish. Rose thinks each column consisted of a few pairs of male and female cod that had found and courted each other in the dense school, and were now rising out of it in order to couple in private.

In 1991 Rose watched a school that had finished spawning

do an equally remarkable thing. The school began to move with a purpose across the continental shelf and toward the northeast shore of Newfoundland. The older and larger fish—Rose calls them scouts—led the way. But as the mob gathered steam, juvenile fish that had been too young to spawn streamed in from all sides and fell in behind their elders. Presently the fish began to spread out, horizontally and vertically, until they were on average eight or ten body lengths apart—just about as far apart as they could get, Rose calculates, and still see their neighbors. Rose thinks the cod were beating the ocean for prey.

While he was watching on his sonar screen, they found it: a school of capelin off to one side but also headed inshore. The cod scouts veered to intercept their dinner; the whole troop, right back to the youths in the rear, followed suit. The capelin rearguard, sensing danger, rose off the bottom in a great cloud, fleeing for their lives. But the cod scouts fell in among them like wolves among sheep, like fighter planes coming out of the sun—their mottled skin camouflaging them against the graveled ground—and tens of thousands of capelin must have come to desperate thrashing ends in those few minutes. All this drama is contained in a few smudges on an echogram if you know how to read it; and all these details of cod behavior have been unknown, hidden by the murky Atlantic, until just recently, even though we have been studying cod for more than a century and eating them for much longer.

Rose was not surprised to find cod migrating en masse. "Fishermen have known for hundreds of years that the cod just don't show up at random," he says. "They show up in particular places, and they all show up at once. So it stands to reason they're migrating in groups. It's like when the robins arrive in spring, or the geese, you know they're not coming one by one. But while we can see the geese flying overhead, we can't see the cod." Rose had an idea of where to look for them, though. As the cod head toward shallow inshore waters in search of capelin, they are also heading toward the freezing glacial melt of the Labrador Current. And so they cross the continental shelf along the bottom of a trench—probably an ancient river-bed once, but now a cod highway. That allows them to stay as deep as possible as long as possible, in the warm (36 or 37 degrees Fahrenheit), salty, and therefore heavy water of the deep Atlantic, which laps through the trench along its bottom.

The migrating cod leave behind their spawn: billions or even trillions of fertilized eggs, transparent spheroids a twentieth of an inch or so across. The buoyant eggs rise toward the surface and float on the current. There the trillions begin to feel nature's winnow: an ocean current is a fluky thing, and it cannot always be relied on to carry a cod egg where a young cod wants to be. In some years and some places, the eggs may get blown right off the continental shelf and out over the abyss. This dooms them. Yet some eggs always make it, and although their route to success off Newfoundland is not well known, on Georges Bank it has been worked out over the past decade by researchers at the National Marine Fisheries Service (NMFS) laboratory in Woods Hole. It is a small marvel of adaptation.

Georges Bank is an oval, about 150 miles long by 75 miles wide, that runs southwest to northeast across the mouth of the Gulf of Maine. As the Labrador Current sweeps into the gulf from the north, it follows the coast around until it flows north-east along the landward edge of the bank. Meanwhile strong tides surge in and out of the gulf across the bank. The com-bination creates a clockwise flow around the bank's perime-ter—a flow the cod take advantage of. They spawn all over the bank, but they are especially partial to the northeastern end. In late winter and early spring, countless cod eggs rise toward the surface and get caught up in the clockwise flow. The lar-vae, bug-eyed and front-heavy, hatch within a couple of weeks. They proceed to grow a mouth, a gut, and an anus; fins, starting with the all-important tail; and finally, scales and teeth. All the while they are drifting southwest along the seaward side of the bank and then northeast along the landward side. And all the while they are getting heavier and sinking lower in the water column.

By August, if all goes well, the larvae have made a partial or even complete circuit of the bank and are ready to settle on the gravelly ground. Greg Lough of the NMFS laboratory has seen them there, through the windows of a submersible, two inches long and stemming the strong tidal current like trout in a stream. The young cod stay out of the worst of the current by hugging the bottom, beating their tail fins furiously to maintain their position. But when a piece of food drifts by—a shrimp, say—the fish pop up and grab it, and let the current carry them a ways before settling back to the bottom. And when a bunch of biologists come by in a submersible and turn on the floodlights, the fish respond in the same way: they pop up into the current and drift off into the black.

Around 99.9 percent of the larvae, though, never settle at all. They may get caught in water that leaves the bank for the open Atlantic, or that fails to make the turn at the southwestern end and instead heads off toward the purgatory of Long Island. They may get descended on by a school of hungry mackerel. Fisheries biologists do not know what happens to the cod larvae that do not make it on Georges Bank—and still less off Newfoundland, where the life history of cod is still full of gaps. Nor do they do know why in some years many more larvae survive than in others.

What they do know is that the difference between a good year and a bad year did not use to matter as much as it does now. A population of fish that live 25 years and lay several million eggs a year each for most of those years is in no danger from the occasional environmental insult. If any year's larvae prove a flop, there are 24 other year classes in the water to pick up the slack. Nor is it only chance environmental fluctuations that cod have proved themselves immune to. Over millions of years, they have repeatedly endured the slow obliteration of their habitat by advancing or retreating glaciers and have still thrived.

And yet in one way cod are not perfectly adapted to their current environment. They have a fatal flaw: firm white flesh, free of oil and bones, which is easily preserved by salting or freezing and which makes an admirable substrate in all sorts of recipes. For helping to feed the expansion of European civilization, *Gadus morhua* is now paying a heavy price.

Cod fishing by Europeans in the New World began in the fifteenth century. The first fishermen to cross the Atlantic were probably Basques, but the Basques left no written records. Europe's first documented encounter with the rich fishery of the New World occurred in 1497, five years after Columbus's first voyage. That is when another Italian sailor, this one working for the English crown, stumbled on Newfoundland. Or maybe it was Nova Scotia; it is not too clear where Giovanni Caboto, John Cabot to us, landed. But it is clear he saw many fish. The evidence is a famous letter to the Duke of Milan from an Italian traveler in England. "They affirm that that sea is covered with fishes," wrote the traveler, recounting a conversation he had had with Cabot and his men on their return,

> which are caught not only with the net but with baskets, a stone being tied to them in order that the baskets may sink in the water. And this I heard the said Master John relate. And the aforesaid Englishmen, his comrades, say that they will bring so many fishes that this kingdom will no longer have need of Iceland. . . .

English vessels had been fishing off Iceland for "stockfish," as cod were then known, since before the fifteenth century. Men in the Iceland fishery were willing to brave the stormy North Atlantic in small open boats because it paid them to; and it paid them to because eating cod was a duty to God. The

Catholic Church required its flock to forgo meat on Fridays and other days of abstinence and during the 40 days of Lent. Even after Henry VIII broke with Rome and started his own church, the demand for fish in England did not flag for long. In 1548, during the reign of Henry's son Edward VI, eating fish was made a civil law, in part to encourage full employment in the fishery. "Considering that due and godly abstinence is a mean to virtue," read the act in its preamble,

> and to subdue men's bodies to their soul and spirit; and considering also that fishers may thereby the rather be set on work, whereby much flesh shall be saved and increased; and also for divers other considerations, it was enacted . . . that none eat flesh on Fridays, Saturdays, . . . nor in Lent, nor on any such other day as is and has been accounted a fish day. . . .

Illicit meat-eaters were fined 10 shillings and locked up for 10 days; repeat offenders got 20 and 20. That same year, to further encourage fishing, England also prohibited the Royal Navy from shaking down fishermen off Newfoundland and Iceland for protection money.

In Catholic France no such royal interventions in favor of fish eating were necessary, and it was France more than England that initially developed the Newfoundland fishery. By the mid–sixteenth century at least 150 French boats were crossing the Atlantic every year. But it was an English ship that expanded the New World fishery to the south. In May 1602 the *Concord* arrived off Massachusetts, searching for sassafras trees—sassafras extract being at the time a popular therapy for syphilis. Anchoring off a "mighty headland," Bartholomew Gosnold and his crew fished for food and caught cod aplenty. They named the headland Cape Cod, although they found sassafras there too. Two decades later the Pilgrims were surviving on cod in nearby Plymouth, and cod was on its way to the special kind of sacredness it would later enjoy in Massachusetts.

In Newfoundland, meanwhile, the English and French had embarked on a centuries-long struggle for cod-fishing supremacy, in which the French were dominant at first but then gradually gave up ground. In 1783, the Treaty of Paris that ended the American Revolution also guaranteed American fishermen the right to fish off Newfoundland. John Adams, one of the negotiators and a Massachusetts man, had threatened to scuttle the whole deal if the British did not concede that right. By the nineteenth century, American "bankers" sailing mostly from Gloucester had come to dominate the Grand Banks fishery along with the French. Both countries subsidized the capital-intensive offshore fishery. The English settlers in Newfoundland stuck mostly to the bays and inlets, letting the cod come to them in spring and summer, when masses of capelin swim right up onto the beaches to spawn and the cod chase them.

The technology of New World cod fishing changed not at all during the first three centuries after Cabot's voyage, and little during the next. It was more or less like sport fishing today, but without the rods and in worse weather. Men stood or sat in a boat and dropped hooked lines over the side. (Strong drink was prohibited after 1743 though, at least on French ships, because it promoted mutinies.) French fishermen on the Grand Banks stood in barrels lashed to the deck to protect them from the weather and wore leather aprons up to their necks. Each man had eight or twelve hemp lines, 500 feet long or so, with a lead weight and an iron hook; and each might catch a hundred cod a day, sometimes as many as 400. Other men beheaded, split, and cleaned the cod and tossed them into the hold, where they were salted. With enough salt the fish could be preserved for the journey back to France. Alternatively, one could fish from shore in a small boat and return to shore that same day to dry the fish. In that case one needed less salt but more land—the split fish were spread out on the beach or on wooden stages. The English, who had no salt of their own and had to buy it from other countries (such as France), favored this mode of operation. That is one reason they rather than the French became the Newfoundlanders.

In the nineteenth century the American, French, and Portuguese bankers began to fish from dories, fanning out from their schooners at dawn, or what passed for dawn in the perpetual fog on the banks, and coming back as early as noon with a boatload of fish. During the summer season the Grand Bank became a small offshore city, as Kipling described it, with hundreds of men in dories working well out of sight of land but often within sight of one another, or at least within earshot of the bells and conches that were everyone's lifeline in the fog. By then the dorymen had begun to give up handlines for longlines: mile-long ropes with hundreds of baited hooks that were set on the bottom for hours and then reeled in. Some inshore fishermen, meanwhile, were switching to cod traps. These were walls of netting that ran out to sea from the land and diverted cod into a boxlike enclosure as the fish swam along shore.

With these simple tools it was possible to catch an impressive amount of fish, as the 60,000-ton harvest from New England in 1895 proved, and over the first four centuries the North American fisheries slowly expanded. Yet North American cod emerged from the nineteenth century relatively unscathed. Fishermen still lacked the tools to wipe cod out; there were still places to which the fish could retreat. "I believe that the cod fishery . . . and probably all the great sea fisheries are inexhaustible," wrote Thomas Huxley, one of the most eminent of nineteenth-century biologists, "that is to say, that nothing we do seriously affects the number of fish." Even in 1883, when Huxley made that remark, it was not true for fish in general; halibut, for instance, were already in decline off the coast of Massachusetts, after suffering a vogue of only a few decades on the dinner tables of Boston. And in the late twentieth century Huxley's claim was to become preposterously untrue even for cod. By then, fishing technology had changed a lot.

The most important change came from Huxley's England. The steam-powered otter trawler was already in use there even as the New England schooners were hauling in their record harvest in 1895. Today the otter trawl is used everywhere groundfish are caught. It is a 150-or 200-foot long net, shaped a bit like a windsock. Giant wooden planks called otter boards are attached to both sides of the net's mouth. As it is towed over the seafloor, scraping the bottom, the force of the water on the otter boards keeps the net open, and fish are swept into the closed end, which is called the cod end—and not specifically because it catches cod. Cod was a Middle English word for bag, and apparently a contemptuous way of referring to a very common fish; later it became a doubly apt name for the bag that caught the fish. (Cod was also a Middle English vulgarism for scrotum, hence the term "codpiece.")

The first American steamer equipped with an otter trawl sailed out of Boston in 1905—to the alarm, not surprisingly, of traditional fishermen. By 1914 the U.S. Commissioner of Fisheries had appointed a committee to investigate what damage the otter trawl might do to fish stocks. The committee reported that the new technology was already causing a decline of fish in the North Sea—not yet of cod, perhaps, those prodigious reproducers, but certainly of plaice and haddock. Proof came in 1919, when English fishermen returned to the North Sea in strength after World War I; they found that their daily catch had more than doubled since 1913. The break from fishing had allowed the fish stocks to recover. Yet neither England nor the United States was moved to impose restrictions on otter trawling.

The otter trawl proceeded to decimate the haddock stock on Georges Bank. Although haddock is a close relative of cod and similar in taste, the demand for it in the United States vastly outstripped that for cod after World War I, especially after packaged, frozen fillets were invented in the 1920s. They made a nice change from leathery salt cod. New England fishermen raced to meet the demand; in 1929 they landed 120,000 tons of haddock from Georges Bank. On a graph showing the haddock catch throughout the twentieth century, 1929 is a hideous fever spike—a warning that was never heeded. By 1934 the haddock catch had plunged to 28,000 tons. After that it picked up a bit until the 1960s, when the haddock on Georges Bank were obliterated again.

During the haddock heyday cod enjoyed a period of obscurity; in 1953 only 8,100 tons were taken off Georges, a record low. In the Newfoundland cod fishery too, the first half of the twentieth century was a period of relative quiet. French and Portuguese otter trawlers began to work the Grand Banks, while the Newfoundlanders continued to fish the inshore in traditional ways (although many of them now had motors on their dories). The catch continued to increase, but it did not skyrocket until after 1954. That is when the British ship *Fairtry* appeared on the horizon.

The *Fairtry* was a new kind of fishing boat: a factory trawler. It was 280 feet long and displaced 2,600 tons, which made it several times larger than the largest trawler of the day, but much smaller than some of the ships that were soon to follow it. As William Warner tells the story in his book *Distant Water*, the *Fairtry* had been commissioned by a Scottish shipping and whaling firm that was keen to expand out of factory whaling now that whale stocks were in decline. Its maiden voyage in pursuit of cod on the Grand Banks was a success; the biggest problem was that the huge net was sometimes filled with so many tons of fish that, when they were hauled in over the stern all at once, the gear sometimes gave way under the strain. "Nothing is more provoking," the captain of the *Fairtry* wrote in one of his reports, "than to see this happen or the cod end burst and the sea covered with dying fish."

By the 1960s, the *Fairtry* had been joined on the North American banks by many more factory ships, not from North America but from the Soviet Union, Germany, and other nations. The largest displaced some 8,000 tons. All of them were designed to fillet and freeze the fish immediately; and all of them were designed to catch enormous amounts of fish. In an hour, one factory ship could haul in as much cod, around a hundred tons, as a typical fishing boat of the sixteenth century could land in a season. In 1968, 810,000 tons of cod were caught off Labrador and on the northern Grand Banks—nearly three times more than had ever been caught in a single year before the advent of the factory ships.

The result, in retrospect, seems entirely predictable. It may even seem astonishing that Americans and Canadians let it happen: that in the space of two decades they let foreign ships all but wipe out of one of their great natural resources. But the freedom of the high seas was a tradition that was not easily jettisoned. Moreover, except among fisheries scientists, and even among some of them, the old prejudice of an inexhaustible sea still held sway. No one thought you could overfish cod on the banks. By the mid-1970s, though, it was clear something had happened. The cod catch had plummeted to less than 200,000 tons off Newfoundland and to less than 30,000 tons on Georges Bank. The haddock on Georges were essentially gone. In 1977, with their own fishermen screaming for help, both Canada and the United States extended their territorial waters out to the present limit, 200 miles offshore. The Exclusive Economic Zone was designed to exclude foreign vessels from most of the fishing grounds.

What is truly astonishing is what happened next. With the factory ships gone, both Canada and the United States had a chance to recreate a sustainable cod fishery. Neither country did. And the fact that biologists were still just getting to know cod—and still learning how to count their far from inexhaustible numbers—was to play an important role in that failure, particularly in Newfoundland.

———

"The basic idea is, once you've eaten all the fish, you know how many there used to be," says Ransom Myers, a population biologist and fisheries expert at Dalhousie University in Nova Scotia. That is how Myers explains "virtual population analysis," which is the state of the art in fish counting. The state of the art in fish counting is not terribly good. Imagine if the only information demographers had about the U.S. population was the number of murders committed each year and the results of the occasional Gallup poll. They probably would have a hard time estimating how many living Americans there are. So it is with fish-stock assessors: they have no direct information on how many fish are born each year, nor on how many die of natural causes. The only fish they can count are the dead ones on deck or dock. Yet they must decide how many fish there are in the ocean, so that policy makers may decide how many may be caught without causing the population to collapse.

Virtual population analysis, or VPA, is the stock assessors' solution to this dilemma. It works like this. Although researchers cannot directly measure how many cod are born in a given year, they can track the progress of that year class once its members are large enough to begin showing up in fishermen's nets. They do so by taking a small but representative sample of the catch as it is unloaded at port. Back in the laboratory, they determine the age of each fish by dissecting out a tiny ear bone, called an otolith, that has annual growth rings like those of a tree (albeit microscopic ones). After a year's worth of sampling, dissecting, and ring counting, biologists can estimate how many of the cod that fishermen caught were three-year-olds, four-year-olds, and so on.

If they repeat this procedure year after year, they will eventually reach a point where no more cod born in a given year are showing up on the dock, because all of them are dead already. Some of those fish never did appear on the dock, because they were eaten by seals or died naturally before fishermen could catch them. Researchers take an educated guess at what the natural mortality rate of cod might be, typically putting it at around 20 percent. Adding that to the fishing mortality— the percentage of the cod population caught each year by fishermen—and adding all the years together, they can count all the cod that were born in, say, 1984, and are now dead.

That alone does them little good, of course. What they want to know is how many fish of all year classes are living in the ocean now, still waiting to be caught. The VPA's accurate census of the dead, though, allows researchers to calibrate their less accurate sources of information on the living. There are two such sources. Each is lousy in its own way.

The first source of information is research surveys—the fisheries equivalent of Gallup polls. The NMFS in Woods Hole and the Canadian DFO in St. John's do these every year. Unlike fishermen, the poll takers do not go looking for fish; they go to hundreds of randomly selected points, trawl, and see what they get. By repeating the same procedure every year with the same gear, they can track changes in a population. The survey does not give them an absolute head count, however, because they have no way of knowing how complete their sample is—how many fish are escaping their nets. The retrospective data of the VPA provide a clue. Since all the cod that were in the ocean in 1984 are now caten and accounted for, researchers do know, belatedly, what relation the research survey from that year bore to the real world. Assuming they did the survey the same way in 1994, that gives them an idea of what the real 1994 numbers are.

A second way of getting a fix on the living fish population

is to analyze the commercial fish catch—or rather, how hard fishermen had to work to catch it. The fewer fish there are in the sea, the logic goes, the longer it takes to catch a given number. By keeping track, year after year, of how many fish fishermen catch for each day at sea, one can chart changes in the fish stock. With the help of the VPA one can then translate that information into an assessment of the current stock. Both methods of counting fish—research surveys and the commercial catch per unit effort—were used in Newfoundland.

Myers and his Dalhousie colleague Jeffrey Hutchings devoted several years of their lives, while both were still working at the DFO in St. John's, to deconstructing the disaster that resulted. By second-guessing those of their colleagues who actually had the responsibility to count cod, they made themselves gadflies within the DFO. "You can't live here in Newfoundland and not be touched by what's been happening," Hutchings said in 1994, after cod fishing on the Grand Banks had stopped. "My family is seven generations here. The fishery is everything. That's it. There's almost no industry, there's no other natural resources except for some pulp and paper, a little bit of mining. It's always been the fishery, and everything is linked to the fishery. So this kind of work is very different from other scientific work—if we get it wrong, this is going to affect a lot of people." In Newfoundland in the 1980s, the cod-stock assessors had gotten it badly wrong.

After the foreigners were kicked off the Newfoundland banks, a brief period of euphoria descended on the province. This was to be Newfoundland's chance at last. Newfoundlanders were still poor, still unemployed, and still living, some of them, in "outports" that could not be reached by road. Past efforts to diversify the economy had more or less failed. Maybe the path to the twentieth century lay with cod after all. The market for salt cod was not what it used to be, but the market for fresh and frozen fish was booming, particularly in the United States, where health-conscious consumers were eating more fish than ever. The foreigners had shown just how many fish could be caught on the offshore banks; now those fish would belong to Newfoundland. The government deliberately set out to encourage the expansion of the offshore fishery, even going so far at one point as to buy a major trawling and fish-processing company itself.

In the midst of this glee, DFO biologists were asked to forecast how fast the cod stock would grow, now that it was no longer being gobbled up by foreign factory trawlers. The stock assessors must have felt considerable pressure; the euphoria must have affected them too. Certainly they fed it. To predict the growth of the cod stock, they had to estimate how many young fish would be "recruited" to the fishery each year. They decided to assume that future recruitment, after 1977, would be the same as the average recruitment during the 1960s and 1970s. During all that time the stock had been continuously declining, as factory trawlers mowed the banks; by 1977, Hutchings and Myers's analysis of the data show, the number of spawning cod off Newfoundland was down by 94 percent from what it had been in 1962. Thus by averaging all those years together, the DFO was being decidedly optimistic.

In particular, it was assuming that the low number of spawning cod would not diminish the size of the newly recruited year class—that the number of mothers, in other words, would not be correlated with the number of babies. To an outsider, this assumption sounds counterintuitive. It is not quite as crazy as it sounds when applied to a species like cod, in which each mother lays several million eggs and the vast majority of them die. Conceivably the number of eggs that make it might depend only on an environmental fluke—how cold the water was that year, for example—and not at all on the number of eggs that were laid. It is an indication of how little we know of cod that so basic a question should still be open for discussion; some biologists continue to assert that there is little relation between cod recruitment and the size of the spawning stock. At any rate the DFO stock assessors believed it in the late 1970s, and they predicted that the diminished stock would rebound rapidly. They said that the cod catch, which had fallen to 139,000 tons by 1978, could safely be increased to 350,000 tons by 1985—and all of that for Newfoundland.

For a few years the DFO got lucky. By chance, the years from 1978 through 1981 did turn out to be ones of relatively good recruitment—although not as good as the DFO had predicted. During those years, with the foreign trawlers gone, the number of cod off Newfoundland increased. But so did the ability of Newfoundlanders to catch cod. Egged on by federal subsidies and by the DFO's optimistic projections, the industry was expanding. Newfoundlanders had otter trawlers now—not factory ships, to be sure, but ships that were large enough to enable them to fish offshore even in bad weather. And those ships had modern navigational and sonar equipment that enabled fishermen to find cod wherever they lived—to locate the large spawning schools and to sweep them up. With the goal of a 350,000-ton harvest in mind, the DFO increased its catch quota on cod several times during the early 1980s. But the fish were simply not there in the numbers DFO was claiming. The agency was cheerleading rather than regulating.

Although fishermen were not even catching as many cod as the DFO was allowing them to, Hutchings and Myers later found, they were catching more than the cod population could sustain. The DFO had started with an overly optimistic forecast of how the cod stock would grow, and from then on it consistently overestimated the size of the stock. Since it had only started doing random research surveys in 1978, it relied at first on the commercial fishing data—the catch per unit effort. The problem with doing that was that fishermen can increase their catch rate when the fish population is flat, and even when it is going down. "You have smart people working very hard to maximize their income by catching fish," Myers explains. "So their efficiency is increasing over time. The stock assessment people thought the cod population was increasing remarkably; they thought fishing mortality was low. In fact, Canadians were basically learning how to fish offshore, and the trawlers were getting more efficient."

By the mid-1980s the problem was apparent, to outside critics anyway, in the DFO's own data. Once most of the fish from the late 70s and early 80s were dead, the VPA revealed that there had been many fewer of them than the DFO had claimed at the time. That meant that fishermen had been harvesting a much higher percentage of the cod population than the DFO had thought—so high a percentage that the population could not possibly be growing at the rate the DFO had projected. Indeed, by 1985, Hutchings and Myers believe, the cod stock was not growing at all. It had begun its long slide into catastrophe.

For the next three years the DFO continued to assert that the stock was growing. By then the stock assessors had started to rely more heavily on their research surveys. As it turned out, that did not help matters. All polls produce highly variable results, but that is especially true of polls of fish. For reasons that no one understands even now, the 1986 survey spiked: it suggested there was an uncommonly large amount of cod off Newfoundland. This encouraged the DFO. Not until the next two research surveys produced sharply lower results did the agency suddenly sit up and take notice.

By then it was too late. Nowhere in the world do scientists dictate how many fish are to be caught; that decision is ultimately a political one. By 1989, the political and economic momentum behind an expanded Newfoundland fishery was too great. Not wanting to throw thousands of people out of work, the Minister of Fisheries and Oceans rejected his stock assessors' shocking advice to cut the 1989 cod quota all at once by more than half, to 125,000 tons. Instead he cut it only to 235,000 tons. "Politically it's very difficult to reduce fishing," says Myers. "Particularly because there's uncertainty—you're causing great economic hardship, and you're not certain. There are a lot of possible errors. But in fact the error was in the other direction—there were even fewer fish and higher mortality than what DFO thought."

Even after its change of heart in 1989, the DFO continued to overestimate the size of the cod stock. "The research surveys appear to have been too high," says Myers. "But they looked OK, and the commercial catch rate data were not examined closely. They were not very clear, but they did show a decline. The analysis of them was completely botched. So you were already taking out too many fish, but because of the error you were taking out tremendously too many."

"And toward the end, as the cod population declined, people tried to maintain their catch rates to maintain their income. So they fished harder. Inshore fishermen were going in small boats 100 miles offshore and setting bottom gillnets—they were going way the hell out in the ocean under incredibly dangerous conditions. So the fish population started to go down more quickly, which caused the fishermen to fish harder.

"Meanwhile it was clear that there were not many older fish, the spawning fish. And in fact there weren't that many younger fish coming through either. There were· lots of people who thought that fishing mortality was too high and that it should be reduced. But no one suspected the magnitude of what was happening. Instead of being two and a half times what was desirable, the mortality was five times too high. And with fishing mortality that high, the stock can collapse very quickly."

In 1991, by Myers and Hutchings's reckoning, Newfoundland fishermen caught more than half the cod living in their waters, some 180,000 tons. In December of that year, the DFO recommended that they be allowed to catch the same amount in 1992.

But it never came to that: in July of 1992 the minister was

forced to close the cod fishery entirely. By then there were next to no cod of spawning age, seven or older, left. There were just 22,000 tons worth, less than a quarter of what there had been in 1977, after the factory trawlers had done their worst, and around an eightieth of what there had been in 1962. Recent history seems to have confirmed that the DFO's initial assumption—that there can be plenty of cod babies even when there are few cod mothers—was wishful thinking. By the end of 1997 the cod stock had shown few signs of recovery, and the moratorium remained in effect for all but a small part of the fishery off the south coast of Newfoundland. It had thrown 30,00 people out of work in the province, out of a total population of 570,000.

The people who were responsible for assessing and managing the Newfoundland cod stock do not see history in the same way that Myers and Hutchings do. They believe something mysterious happened in 1991—"a sudden, drastic and unexpected" drop in cod, as a couple of senior DFO biologists referred to it in their own postmortem paper. Maybe the water got too cold for cod; maybe harp seals ate them. "Some people believe there was actually a huge number of fish out there," says Myers. "Then all of a sudden they died or something, and no bodies were found. And it's not our fault, it's the environment—the environment changed. I think that's total nonsense."

For saying as much to a Canadian newspaper reporter, Myers was officially reprimanded by the DFO. In 1997 he left the department, following Hutchings to Dalhousie. Meanwhile the DFO had decided to subsidize seal hunting; in 1996 it raised the harp seal quota to 250,000 animals year, partly out of concern that the swelling seal population was preventing a comeback of the cod. With so few cod left in the water, it was not an unreasonable hypothesis. But there was not much in the way of data to back it up.

New England had two years to learn from Newfoundland, and it learned nothing. The stakes were always lower there. Georges Bank and the Gulf of Maine were never anywhere near as rich in cod as the Grand Banks; Gloucestermen used to sail to the Grand, after all. Moreover, it has been a long time since fishing was as dominant a component of the economy in New England as it is, or was, in Newfoundland. But it still matters a lot in places like Gloucester and New Bedford. The long history of the cod fishery should have been reason enough not to throw it away.

After 1977, though, with the creation of the 200-mile limit, the New England fishing industry experienced the same euphoria as Newfoundland. New England fishermen had lobbied Congress hard to have the foreign trawlers kicked out, and they expected a bonanza. Between 1977 and 1983, the number of boats fishing out of New England ports increased from 825 to 1,423. The new boats were bigger and equipped with the latest electronic fish-finding equipment. As in Newfoundland, the fish never had a chance. The cod catch on Georges Bank alone peaked in 1982 at more than 53,000 tons. Then it started to decline. As the stock declined, the mortality inflicted by fishing rose, just as it did in Newfoundland. The difference is that in New England, fisheries biologists knew it was happening all along and said so.

Under the Magnuson Fishery Conservation and Management Act of 1976, the National Marine Fisheries Service was charged with assessing the status of the various fish stocks and with overseeing their management. But recommendations on what restrictions, if any, to place on fishing were left to eight regional councils composed mostly of fishing industry representatives—fish processors, fishermen's association leaders, and fishermen themselves. During the 1980s, the New England council proved itself unwilling or unable to control fishing. Indeed, one of its early actions, in 1982, was to eliminate catch quotas. Its goal, it said, was a simpler system of regulations that would allow the fishery "to operate in response to its own internal forces. . . ."

As the decade progressed, the fishery did just that; and as NMFS scientists warned of declining stocks of cod, haddock, and yellowtail flounder, the council took no effective action. That changed only in 1991, when a suit by the Conservation Law Foundation forced the council's hand. Under court order, it began considering regulations that would require fishermen to reduce the numbers of days they spend at sea by 10 percent a year for five years—the goal being to cut fishing effort and presumably fishing mortality in half. Fishermen protested; while the council system lets the foxes into the hen house, it manages to leave most of them feeling unrepresented. Nevertheless, the new regulations, known as Amendment 5, took effect in May 1994.

Three months later the NMFS announced that those restrictions would not be nearly enough to save Georges Bank cod—let alone haddock or yellowtail flounder, which had already collapsed. In 1993, the NMFS survey had found, fishermen had caught 55 percent of the cod living on the bank. Only reducing fishing immediately "to levels approaching zero" could save the stock now. In December 1994, the NMFS implemented emergency regulations to close large parts of Georges Bank—the parts where the last cod were huddling—to all ground fishing, while the New England council tried to figure out a long-term solution. That was what the council meeting in the Peabody Holiday Inn in late 1994 was about: it was about picking up the pieces.

In the audience that day was a NMFS scientist named Andrew Rosenberg, who had recently been named to head the fisheries management division in Gloucester. His experience of the New England situation was longer though; in the early 1990s he had been the NMFS liaison to the fishery council, charged with passing on the scientists' advice. During the lunch break of the day-long meeting Rosenberg reflected on the frustrating failure he had been a witness too. "When the Magnuson Act came in," he said, "its primary aim was to Americanize the fishery. It had all the tools in it to do more than that. It had the tools to put in strong regulations. But in New England, the attitude was, 'Now we got the foreigners out, it's our turn.' And so there was a very rapid buildup of fishing effort. So now it's come to the point where unless you're absolutely blind you can't pretend that the stock isn't in very bad shape, because it's almost gone. And still people are arguing 'We don't want to have a direct control on how much we catch.' "

The dreadful irony of overfishing, as Rosenberg or any other fishery scientist will explain, is that if it could somehow be stopped, and fish stocks were allowed to grow again, and fishermen fished at a rate that was lower than the stocks' growth rate, they could catch more fish with less effort. In New England, NMFS economist Steven Edwards and population biologist Steven Murawski have estimated, overfishing costs the economy $150 million a year in groundfish that never got a chance to be born, let alone caught. For the United States as a whole that loss has been put at approximately $2 billion a year. "It's just like interest in a bank account," said Rosenberg. "If your bank account is earning 5 percent interest, and you take 10 percent out every year, what happens to the account? It drops like a stone. If you take 3 percent of the account every year and it's earning 5 percent interest, it grows. And eventually 3 percent of a big number is bigger than 10 percent of a very small number.

"When it was first discussed," he went on, "Amendment 5 probably could have done something for cod. You had some big year classes in the late '80s which could have helped production. If some protection had come in rapidly at that point, if somebody had said, 'Hold it, we've got something coming through, instead of just fishing it harder, let's really back off and let that build a new stock for us'—if they had done that, then you would have been in relatively good shape. But by the time Amendment 5 came along those good year classes were gone. It's the bank account analogy again. If you've got your money in the bank and suddenly you get a little inheritance, you can either blow it all in the first year, or you can hang onto it. After time, 5 percent of the inheritance will be quite a large number, and you'll be able to use it for a long period. The option chosen here was essentially equivalent to blowing it."

What is to be done? The few cod that are left are doing their best already—still chasing capelin and herring, still stemming the currents on the banks, still ascending by twos from their diminished schools to couple gently in the gray unquiet Atlantic. On Georges Bank, at least, they are responding just as biologists would expect to a predator that is slaughtering them: they are spawning sooner—at age two now instead of age three. They are living faster because they are dying younger. As their numbers have declined, they have had to surrender their dominance on the bank to other species, such as skates and spiny dogfish. These fish are less attractive to consumers (although fishermen are going after them nonetheless), and they prey on young cod, as harp seals do off Newfoundland. How this will affect the recovery of the cod stock, no one knows. By the beginning of 1998, there were some signs that the stock was increasing, but only because key areas of Georges Bank were still closed to all fishing. Meanwhile, the cod stock in the Gulf of Maine, where fishing continued, was on the verge of collapse. Off Newfoundland, Myers predicts that the recovery will take another 20 years.

Overfishing does not only wipe out individual stocks; it can change whole ecosystems. A 1998 report by fisheries scientists in British Columbia and the Philippines dramatized the extent to which that has become a global problem. More than half the world's commercially valuable fish stocks are threatened by overfishing, if they have not already collapsed like cod. That was already well known. But Daniel Pauly of the University of British Columbia and his colleagues revealed a different trend: the tendency to "fish down the food web" and catch smaller, plankton-eating animals, such as capelin, herring, or even krill, in place of higher-level predators like cod. Analyzing global statistics compiled by the Food and Agricultural Organization between 1950 and 1994, the researchers concluded that there had been a gradual but steady decline in the average rank in the food web of the fish that are caught. "If things go unchecked, we might end up with a marine junkyard dominated by plankton," Pauly told *Science*. That would certainly make it more difficult to rebuild collapsed stocks of the larger fish.

Consumers in North America have not suffered much yet from the Atlantic cod collapse; for the moment, the slack is being taken up by such alternatives as Pacific cod and pollack, which is now being swept up by factory ships off Alaska. As taxpayers, though, they have had to pay for the mismanagement of the fisheries. The U.S. government has allocated more than $60 million to various programs designed to ease the pain of New England fishermen, including a program to buy their boats, that is, to pay them not to fish. Canada, with a bigger problem and a long tradition of compensating unemployed or seasonally employed fishermen, has spent more than $1 billion. Clearly, when a public resource is being given to a group of people, who are then allowed to ruin it, and who must then be paid not to ruin it, a deep irrationality is at work. On that much most people can agree, but not on what to do about it.

Economists say the problem with current fisheries management is precisely the open access: as long as the resource is free and open to all, it will be valued by none, and overexploitation will be inevitable. Fish are in fact the only natural resource in the United States that is still given away on a large scale; even cattle ranchers and timber companies pay fees, however small, for their use of public lands. To many economists, the solution is to give the fish away or auction them off once and for all: to adopt some form of private ownership, either of the fish themselves or of the right to catch a certain number of them. Those property rights could then be freely bought and sold. Over time the more efficient fishermen—or fishing corporations, more likely—would presumably buy out the less efficient ones, and in theory all that efficiency would be good for consumers. Meanwhile fishermen with an ownership stake in, say, Georges Bank cod, might be expected to husband the resource better than a government would. They might start, one economist has suggested, by weeding out some of those cod-eating dogfish.

Anthropologists who study fisheries are more likely than economists to worry about the effects of too much economic efficiency—about what happens to traditional fishing communities, for instance, when large corporations start buying up the fishing rights. But they too tend to think that fisheries would be managed more wisely if less power were wielded by central government agencies like the DFO and NMFS and more were in the hands of local groups of fishermen. The problem with such an approach in a fishery so diverse as New England is that the fishery is not local, and fishermen there tend not to agree on much. Georges Bank and the Gulf of Maine are fished by Yankees out of Maine, Italians out of Gloucester, and Portuguese out of New Bedford. A few years ago Madeleine Hall-Arber, an anthropologist at M.I.T., made a formal study of this diversity, asking fishermen among other things why they think stocks have declined. "Fishermen tend to blame pollution [and] global warming," she concluded, but also "fishermen using other gear and fishermen from other ethnic groups." At the Peabody meeting in late 1994, a fisherman from Cape Cod got up to recommend, not implausibly, a mandated return to the hook-and-line fishery; a Maine gillnetter moved that otter trawls be banned from the Gulf of Maine; and otter trawlers from Gloucester defended themselves vehemently against all such attempts at "discrimination."

Most fishermen in New England and Newfoundland must recognize that their way of life is in trouble. But most are anything but rich, and many chafe at restraints. Freedom to work when they want and how they want is one of the things that draws people to fishing in the first place; and in an age of proliferating "rights," the "right" to fish is one of the older ones. What the cod crisis demonstrates, unfortunately, is that the world has become too small and our own numbers too large for such a right to be acknowledged anymore. It is a privilege that has been abused. That is hard to accept.

"You're talking about people's livelihoods," says Rosenberg. "A scientist looks at it simplistically: 'The harvest rate should be this, and it's not, so therefore you should reduce it.' But when you reduce it, who goes out of business, and who has to move away, and who is unemployed? The scientific view is very simplistic, and I think that most scientists realize that. They don't expect everybody to bow down to the ground everytime they say 'Well, you ought to do this or that.'

"The difficulty, though, is that at some point there is a biological bottom line. When I was liaison officer to the New England council, I used to call my job 'A Thousand Ways of Saying, You're Killing too Many Fish.' I was supposed to be passing on the scientific advice—and that's what it was: 'You're killing too many fish. Too many fish are dying due to fishing. The fish are dying more rapidly than they're reproducing.' You could say it a thousand ways. But it amounts to the same thing."

All over the world people have failed to hear that message; all over the world, through a combination of desperation and greed, ignorance and mismanagement, people are finding the bottom of fish stocks that once seemed bottomless. Yet it is still shocking that it should happen to cod—stolid, prolific, resilient cod, numberless cod, beef of the sea. It is shocking precisely because we never really did hold cod sacred, not the real flesh-and-blood animals anyway. Though we hardly knew them, we took them for granted—much as hunters once took the buffalo for granted when the prairie was black with them. There is no great mystery about what happened to the buffalo, and none either about what happened to the cod off northeastern America. Men like the ones in that Holiday Inn ballroom—the last of the buffalo hunters—caught them. And the rest of us ate them.

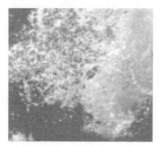

Where the
Water Goes

In our mind's eye we can almost see it whole, the round-the-world journey that seawater takes. We can imagine taking the trip ourselves.

It begins north of Iceland, a hundred miles off the coast of Greenland, say, and on a black winter's night. The west wind has been screaming off the ice cap for days now, driving us to ferocious foaming breakers, sucking every last ounce of heat from us, stealing it for Scandinavia. We are freezing now and spent, and burdened by the only memory we still have of our northward passage through the tropics: a heavy load of salt. It weighs on us now, tempts us to give up, as the harsh cold itself does. Finally comes that night when, so dense and cold we are almost ready to flash into ice, we can no longer resist: We start to sink. Slowly at first, but with gathering speed as more of us join in, and as it becomes clear that there is nothing to catch us—no water underneath that is denser than us. We fall freely through the tranquil dark until we hit bottom, more than a mile and a half down.

There we join a pool of others like us—a pool of cold, salty water that fills the Greenland and Norwegian basins. From time to time the pool overflows the sill of the basins, an undersea ridge that stretches between Greenland and Iceland and Scotland. Then the falling starts again. Only now it is not a parachute drop but a headlong rush, down slope and tumbling like a mountain stream, but far more powerful, more powerful even than Niagara: a giant underwater waterfall, cascading over seafloor lava piles and sediment drifts into the Atlantic abyss. Falling, we pull shallower water in behind us. From our right flank, as we reach the latitude of Newfoundland, we are joined by a cohort from the Labrador Sea, where the winters are brutal too; not quite as dense as we are, this water settles in above us, headed south along the slope of North America. Near Bermuda our ranks are swelled from the left by spinning blobs of warm Mediterranean water, even saltier than we are; they sail like Frisbees out of the Strait of Gibraltar and cross the ocean to join us. Greenland water, Labrador water, Med water—we all fall in together, and gradually we mingle: we are North Atlantic Deep Water now. Mediterranean salt seeps through us like a dye. At every step on the road, some of us lose heart and turn back—back toward the north, up toward the surface. Yet still our mighty host advances, 80 Amazon Rivers marching along the ocean floor, toward the equator and across it.

All through the South Atlantic our army remains intact, hugging the western slope of the ocean basin. But that reassuring guide ends where South America does, and in the stormy Southern Ocean we are scattered by the great centrifuge, the mixmaster, the buzz saw—what metaphor can do justice to the Antarctic Circumpolar Current? Sweeping around the frozen continent from west to east, touching every ocean but never any land to stop it, it carries not 80 but 800 Amazons of water. It blends the waters of the world, obscuring their regional roots. Those of us of North Atlantic origin are dragged to the surface off Antarctica, where we absorb a blast of cold and quickly sink again. We spread north now into all the oceans, mostly at a depth of half a mile or so—some back into the Atlantic, some into the Indian Ocean, and many of us into the Pacific. In that vast and empty basin we drift northward, hugging western slopes where we can, until we reach the equator; there the trade winds part the waters, and tropical heat mixes down into us, buoying us to the surface. It is time to head for home.

Now comes the confusion of Indonesia: South of the Philippines and into the Makassar Strait we funnel, twixt Borneo and Sulawesi, only to slam into a near-impenetrable wall of islands. Some of us blast our way past Bali through the Lombok Strait, impossibly narrow as it is; the rest opt for the eastward detour around Timor. Into the open Indian Ocean at last, we make for Africa, collecting salt from the hot shallows of the Arabian Sea as we once did from the Mediterranean. Southward then down the coast of Mozambique, and we are picking up speed, in preparation for our triumphant return to the home ocean; but rounding the Cape of Good Hope is no easier for water than rounding the Horn is for sailors. Again and again we are beaten back. Only by detaching ourselves in spinning eddies from the main current do some of us manage to sneak, rather ignominiously, into the South Atlantic. There we are joined by water that never bothered with Indonesia and Africa, but instead took the colder shortcut around South America, through the Drake Passage.

One last obstacle remains for us all: the equator, where we must cross the 12-lane highway of east-west surface currents set up by the trade winds. We do it again in eddies, giant ones that spin us north along the Brazilian and Venezuelan coasts and finally shatter in the Caribbean. In the process they dump us into the Gulf Stream at its source off Florida. This is the home stretch at last; Iceland looms ahead. A millennium has passed since we left.

Oceanographers call this global journey the thermohaline circulation, because it is driven primarily by heat and salt— the two properties that change the density of seawater, causing it to rise or sink and spread across the top or bottom of the ocean. The thermohaline circulation is more than a natural curiosity. It spreads solar heat from the tropics to the high latitudes; it is what keeps Europe, for instance, warm and habitable. There is evidence that it has been going on in much its present form since the Miocene, and at least since three million years ago, when the Isthmus of Panama emerged and cut off the Atlantic from the Pacific. Given its antiquity and its tremendous force, one might imagine that nothing short of continental drift could change the thermohaline circulation. And one might dismiss as preposterous the notion that human beings, of all feeble agencies, could affect it at all. But the evidence suggests otherwise. We might already be on our way to shutting it down, with consequences we can only dimly foresee.

The basic scientific principle underlying the thermohaline circulation has been known for two centuries. It was discovered in 1797 by Benjamin Thompson, Count Rumford, who in spite of his title was an American reared in poverty. In 1774, Thompson had been chased out of Rumford (now Concord), New Hampshire, by an angry mob, which suspected him of Tory sympathies—rightly as it turned out—and desired to tar and feather him. Abandoning his wife and baby daughter (he had married for money), Thompson became a British spy until he was forced to flee to London a year later. In London he cultivated the right people and rose to prominence; George III even knighted him. But it was in Bavaria, where he moved in 1784 to serve the local potentate as war minister, that he became a count of the Holy Roman Empire. Graf von Rumford was his choice of title. By all accounts he was not a very nice man, nor one with a well-developed sense of shame.

But Rumford was an inventive man. Among his many credits were the soup crouton, a drip coffeemaker, a kitchen range, and the potato, or rather its popularization in Europe. Clothing and feeding the Bavarian troops was one of Rumford's responsibilities—the crouton was meant to keep them from wolfing

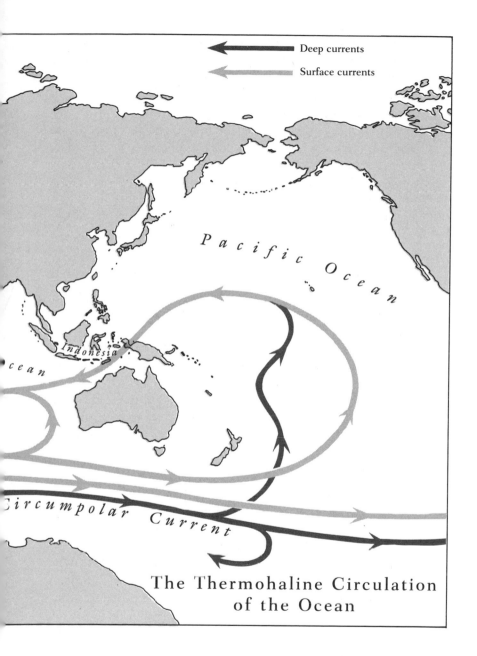

Deep currents

Surface currents

Pacific Ocean

Indonesia

cean

Circumpolar Current

The Thermohaline Circulation
of the Ocean

their watery soup—but there was also a consistent intellectual theme to his work: an interest in heat. One day, after burning his mouth on a piece of apple pie, he resolved to understand the nature of the beast. A chance observation then set him on the path to his great discovery.

After heating an enormous thermometer—its copper bulb was four inches across—Rumford had placed it in a window to cool. The alcohol in the thermometer was dusty, and as sunlight fell through the glass, Rumford noticed something strange: the dust specks were moving—and not just any which way, but in a vigorous column up the center of the thermometer and down again along its sides. Rumford had discovered convection currents. They must have been observed before, by cooks standing over boiling pots of soup if by no one else, but Rumford seems to have been the first to take scientific notice of them. He certainly was the first to grasp how convection currents might operate in the ocean.

"If the water of the ocean, which, on being deprived of a great part of its Heat by cold winds, descends to the bottom of the sea . . . ," Rumford wrote in an essay in 1797, "it will immediately begin to spread on the bottom of the sea, and to flow towards the equator, and this must necessarily produce a current at the surface in an opposite direction; and there are the most indubitable proofs of the existence of both these currents." One of those proofs had been collected by a British slave ship in 1751 off the sweltering coast of West Africa. Lowering a thermometer in a bucket, Captain Henry Ellis of the *Earl of Halifax* had discovered that the water at a depth of a mile was no more than 53 degrees Fahrenheit. (The actual temperature, we now know, is closer to 40 degrees; Ellis's thermometer had been warmed on the way up.) The cold deep water was useful for chilling wine and the captain's bath, which, as Ellis later explained, was "vastly agreeable in this burning climate." Rumford recognized the deeper significance of the observation: there was no way to get water that cold in the tropics without bringing it in from the poles. In the deep ocean, water that had been chilled at the surface in high latitudes had to be flowing toward the equator.

In the century and half after Rumford's seminal essay, oceanographers' understanding of the physics of deep-ocean currents did not get much deeper. Their data archive did, however. The *Challenger*, on her pioneering circumnavigation in the 1870s, was the first to follow Ellis's lead in an extensive and systematic way, taking the temperature of deep water at stations all over the world. For sheer systematism, though, the *Challenger* could not compare with the German *Meteor* expedition; between 1925 and 1927, the *Meteor* steamed back and forth across the Atlantic along 13 latitude lines, spaced at regular intervals from 20 degrees north—just below Cuba—to the tip of South America. (The higher northern latitudes had been fairly well measured by then.) Her crew made more than 9,000 measurements of water temperature and salinity at 310 stations.

Although other research vessels followed the *Meteor*, it was her measurements above all that led to the picture of Atlantic anatomy that still survives today. In this picture, deep water forms in only a few small regions—the regions where surface water gets dense enough to sink. And only in two places does water become dense enough to sink all the way to the bottom: off Greenland and in the Weddell Sea, off the Antarctic Peninsula. The Atlantic is like a bathtub with a faucet at either end, running all the time, more or less, filling it from the bottom up. The two water masses, North Atlantic Deep Water and Antarctic Bottom Water, meet in the South Atlantic, and the latter, being the denser of the two, wins the battle for the bottom. (Its temperature is around 30 degrees, below the freezing point of fresh if not of saltwater.) Antarctic water slides into the North Atlantic along the ocean floor; North Atlantic water rides south and up the sloping front of the colder water mass until it surfaces in the Antarctic. There some of it is immediately chilled and sinks to form a new water mass, Antarctic Intermediate Water, at a depth of about half a mile. Thus does the Atlantic become a layer cake of water masses, all readily distinguishable by their particular combinations of temperature and salinity.

The *Meteor*'s measurements revealed all these layers. But they could suggest only in the crudest way how the layers came to be—how the water was moving through the deep ocean. In the absence of any real physical understanding, oceanographers made the simplest assumption: that the polar water masses were advancing toward the equator in broad fronts, across the whole ocean basin, and at a glacial pace. Indeed, over the years the layers of the ocean came to seem practically geologic in their fixity. The idea that there might be vigorous currents in the abyss was not contemplated at all, not until an oceanographer named Henry Melson Stommel came along.

"It was a bedlam of hot-tempered . . . contentious women. There was no peace among them." Thus Henry Stommel on his childhood in Wilmington, Delaware, and later in Brooklyn, growing up fatherless in the Depression, in the company of a poor and embittered mother, a grandmother, a divorced aunt, and a great-grandmother. "I know from where the fierce demon within me that never lets me rest comes," Stommel wrote, "and why I had to seek a private refuge in my own mind and science." By the time he was prevailed on to produce an autobiographical essay, in the mid-1980s, Stommel was already a legend among oceanographers. No one in this century had made more fundamental contributions to the field; no one had been half so influential in a personal and oracular way. Oceanographers made pilgrimages to see Stommel at his farmhouse on Cape Cod, where he worked for decades at the Woods Hole Oceanographic Institution, and in later years they would vividly remember their brief contacts with him. He was close to few of them but friend to many, and often his friendship expressed itself in intellectual generosity—in passing off ideas for others to work on. Stommel's own demon could not linger long on any but the biggest oceanographic problems.

In considering those his gift was the capacity for simplification: for reducing the real welter of the ocean to a tractable mathematical essence. In his autobiography he touched briefly on the question of where that gift might have come from. "During my high school years I developed a taste for simple ideas interacting with a sparse number of facts: things which could be kept entirely within my head without reading notes or having to write them out," he wrote. "The reason seems to have been the difficulty I had reading with spectacles which had been mis-prescribed." Later, the headaches he got from reading kept him from going into the law, as a counselor at Yale advised him to do (a freshman-year aptitude test had revealed Stommel's ineptitude for science). During World War II, when he declined military service as a conscientious objector, he thought for a time of taking up the ministry. He ended up taking a job at Woods Hole instead, doing research related to antisubmarine warfare. From that it was a short hop, once the war had ended, to a career in physical oceanography.

Bad glasses and an unhappy childhood make an unsatisfying explanation for creative genius—but then a taste for simple ideas does not necessarily encourage insightful autobiography. (Stommel's publisher rejected his effort.) Insightful simplification, though, was just the thing needed to make progress in understanding something as big and fractal as the circulation of the ocean. When Stommel set out to explain the existence of the Gulf Stream, on practically his first crack out of the box as an oceanographer, he imagined the ocean as a rectangular box with a flat bottom.

By then, in 1946, oceanographers had known for decades that surface currents like the Gulf Stream are driven primarily by the winds. In all the ocean basins, trade winds near the equator push water toward the west, while westerly winds in the midlatitudes push it toward the east; the result is great current gyres that spin clockwise in the Northern Hemisphere and counterclockwise in the Southern. Prince Albert of Monaco, an avid oceanographer, had proved the existence of the North Atlantic gyre as early as the 1880s, by means of the message-in-a-bottle technique. From his yacht the *Hirondelle*, he dropped hundreds of bottles and barrels into the eastern Atlantic and found that many of the messages in them were returned from points south. A couple of Albert's floats were even picked up in the West Indies. They had become caught in the North Equatorial Current, the southern leg of the gyre, which feeds water into the Gulf Stream.

The Gulf Stream is not really a stream, then; it is merely the western rim of a giant, spinning lens of water. But—and this was one of the great mysteries about surface currents until Stommel came along—the western rim is not the same as the eastern. The center of the North Atlantic gyre is not in the center of the ocean, equidistant between Europe and North America; it is displaced to the west, into the western Sargasso Sea. To the east of that center, water flows southward in a broad, lazy, barely perceptible drift—the one that Prince Albert discovered. It moves at only half a knot or so. To the west, that same volume of water bores through a much narrower channel, and so it moves faster, as fast as five knots in some places: that is the Gulf Stream. The Gulf Stream is not unique. The same type of intense "western boundary current" exists in other ocean basins as well. The Kuroshio, which flows north along the coast of Japan and Siberia, is one example; the Agulhas Current, which flows south along the east coast of South Africa and attains velocities comparable to the Gulf Stream, is another. Stommel first heard about western boundary currents as a problem worth thinking about from a colleague named Ray Montgomery, on a car trip from Woods Hole to Rhode Island in 1946. As Montgomery later recalled,

Stommel solved the problem qualitatively over coffee at a highway rest stop.

That does not mean his reasoning is easy for the lay person to grasp; indeed, professional oceanographers often have trouble explaining it even today. But at bottom what Stommel did was to figure out how to apply a universal physical law, called the conservation of angular momentum, to the circulation of the ocean. The angular momentum of anything that spins is obtained by multiplying three of its properties: its mass, the square of its radius, and its angular velocity, or rate of spin. According to the conservation law, the angular momentum of a body must remain constant unless that body is being pushed by an external force. The figure skater who pulls in her arms, for instance, decreases her radius, and so for her angular momentum to remain constant, she must spin faster. (If she were to throw up from all that spinning, thereby decreasing her mass, she would also speed up a bit.)

The figure skater's counterpart in the ocean is a vertical column of water. Since it is hard to keep track of the mass and radius of a moving water column in a turbulent ocean, oceanographers find it more easy to say that it is the water column's "vorticity" that is conserved rather than its angular momentum per se. Vorticity can be thought of as the tendency to rotate, and all seawater has it, even water that is not actually moving in circles, but merely along a curving path. It can be a clockwise tendency or a counterclockwise one, and a given water column may have both tendencies at once, even though only one triumphs in the end. In fact, every water column gets vorticity from two sources that sometimes battle one another. First, the wind or a collision with a coastline may set the column spinning on its axis, relative to the rest of Earth's surface. And second, just by virtue of being on a spinning planet the water column partakes of a planetary vorticity, which is counterclockwise in the Northern Hemisphere and clockwise in the Southern. Relative vorticity plus planetary vorticity equals the water column's absolute vorticity—and that, assuming the depth of the column remains unchanged, is what is conserved.

Stommel's key insight was to recognize that planetary vorticity varies with latitude, and that it thereby exerts a big effect on ocean currents. (It is more usual to describe this effect in terms of a fictitious force, the Coriolis force, but being fictitious, that force tends to be even more head spinning than the concept of vorticity.) The planetary vorticity of a water column is greater the more its own spin axis, which is perpendicular to Earth's surface, comes into alignment with Earth's—that is, the closer it gets to one of the poles. It is zero at the equator, where the column's axis and Earth's are at right angles. So when a column of water moves from one latitude to another, its planetary vorticity changes. Because its overall vorticity must remain the same, its relative vorticity must also change— by the same amount as the planetary vorticity but in the opposite direction. The water column spins faster or slower, just like the figure skater who moves her arms.

Now think of the North Atlantic gyre as being composed of innumerable such water columns, all spinning clockwise, like the gyre itself, as they are pushed by the winds. The winds are an external force that increases the angular momentum of the whole system. But clearly the gyre and the water columns in it cannot simply spin faster and faster indefinitely. The gyre we see today has settled into a steady state, in which the force of the winds is balanced by another external force: friction. The faster the current, the greater the friction. As the gyre rubs against the European and American coasts, some of the clockwise vorticity is drained from it; this prevents it from spinning out of control.

As Stommel showed, however, the balance is achieved in very different ways on the east and west sides of the gyre, thanks to vorticity conservation. On the east side, as a water column moves south toward the equator, it loses counterclockwise planetary vorticity. To keep its total vorticity the same, it must lose some of its *clockwise relative vorticity* as well. In other words, the planet's rotation helps strip the water of some of the clockwise spin imparted by the winds. That leaves less of the job to be done by friction, which explains why the current in that half of the ocean can be a broad, slow drift.

But on the west side of the Atlantic—and of other oceans— the situation is just the reverse. As water moves north along the American coast, away from the equator, it *gains* counterclockwise planetary vorticity, and to balance that, it must *gain* clockwise vorticity as well. Here, in other words, the planet is reinforcing the winds and aggravating the problem of maintaining a vorticity balance. The only way for the gyre to lose all that clockwise vorticity is to have a lot of friction, which requires a fast current. If there were none, the law of vorticity conservation and thus of angular momentum conservation would have to be violated. No figure skater, let alone a physicist, could countenance that. That is why there is a Gulf Stream.

Stommel published this theory (although not quite in this form) in a short paper in 1948. Explaining why the most famous ocean current exists, a current that had first been mapped by Benjamin Franklin, was a dramatic way to start a career in physical oceanography, but it was just a start. In the 1950s Stommel turned his attention to deep currents, which, far from being famous, were scarcely believed to exist at all. He found that the same insight that had produced his breakthrough on the Gulf Stream question—the importance of the variation with latitude of the planetary vorticity—yielded a surprising prediction when it was applied to the abyss.

The *Meteor* and other expeditions had established that cold water sinks into the abyss primarily near the two poles, but they had also shown that it is not confined to a thin layer near the ocean floor. On the contrary, it is the ocean's warm water that is confined to a thin layer near the surface. Even in the sun-baked tropics, the warm layer is not more than a thousand feet thick; below that lies a layer, called the thermocline, in which the temperature falls rapidly, such that by a depth of a half-mile or so it is no more than 40 degrees Fahrenheit. Clearly the heat of the sun is not conducted right through the ocean, as one might expect it to be; something props up that thermocline. Stommel decided it must be cold water welling up from the abyss all over the ocean.

That led him to a bizarre conclusion. Rising through the ocean is equivalent to moving toward the equator, in that it takes the water farther from Earth's spin axis and thereby decreases its planetary vorticity. To counteract that effect and keep its vorticity the same, the upwelling water must also be moving *away* from the equator; in the deep and open ocean, far from winds and coasts, there is no other source of spin besides the planet itself. Paradoxical as it sounds, in most of the ocean, cold deep water must be flowing *toward* the poles rather than away from them.

But that requires the cold water to have gotten away from the poles in the first place. Where does it do that? On the map Stommel envisioned of the North Atlantic, cold water was sinking into the deep off Greenland, and across most of the abyss to the south it was curling back toward Greenland in a counterclockwise direction. There was only one way to close that circuit. Water had to be flowing south in a narrow current along the American coast, directly under the Gulf Stream but in the opposite direction. And to supply a whole ocean basin with cold water, that narrow current had to be flowing fast.

This prediction of Stommel's, first made in 1955, is still regarded with awe and reverence by physical oceanographers, because it was just that—a prediction. The Gulf Stream theory was remarkable enough, but it was in a sense conventional oceanography: Stommel had taken a known set of facts and offered an explanation for them. His abyssal-circulation theory, on the other hand, predicted a new set of facts that no one had yet observed. It provided a way of proving itself wrong, which is supposedly the essence of good science but which was new to oceanography. Stommel was sticking his neck out.

He did not have to keep it extended for long. Laboratory studies with rotating tanks soon confirmed that his idea was plausible. Stommel himself then suggested a way of measuring real deep-ocean currents: by making a float that is less compressible than seawater. If the float were denser than water at the surface, it would sink at first, but at some depth it would stop—the depth at which the seawater had been compressed, by all the water above it, to a greater density than the float's. Drifting on the current at that depth, the float would emit sounds that would allow its progress to be tracked from a distance. Stommel imagined that it might detonate small bombs once a week, which would be heard halfway across the ocean.

Even as he was publishing this idea, a British researcher he had never met was already building such a float, but without the bombs. John Swallow was an oceanographer's oceanographer: a man with a pure love of studying the sea. He had discovered his inclination in World War II, when he was drafted into the Royal Navy and, as he later recalled, "found I was quite happy going to sea and crawling around a ship mending electronic equipment." After the war he found he was even happier to go to sea for four years straight, on a circumnavigation aboard a latter-day *Challenger*. All his life—he died in 1994—Swallow would spend an average of a third of the year afloat, even after he was supposedly retired. He recommended it as a way of concentrating the mind and escaping the distractions of daily life, provided one went for at least a month. Swallow was a soft-spoken man, modest about his own insights into questions of great import, oceanographic or otherwise; a man who was quite happy not to be working on anything "important" at all; a man who liked quiet and liked tinkering with instruments—and who had a great gift for making them work at sea.

The instrument he designed to measure deep-ocean currents, and in particular to test Stommel's theory, was built of two 10-foot lengths of aluminum scaffold tubing bound together. One tube was just for flotation; in the other, Swallow crammed the batteries and electrical wiring needed for driving a transducer that protruded from the bottom of the tube. When that nickel ring was set to vibrating at 10,000 cycles per second, the sound could be picked up a few miles away by hydrophones hanging over the side of a research ship. By making several such measurements from different positions, Swallow could get a fix on the position of the float relative to the ship. As it trailed along behind the float, the ship could get a fix on its own position (in those days before satellite navigation) by bouncing a radar beam off a buoy anchored nearby.

None of this was easy, in spite of the way Stommel would summarize the work decades later in his autobiography. "The existence of the deep undercurrent flowing southward was quickly confirmed," he wrote, describing the cruise he and Swallow had gone on from Bermuda to Charleston in March 1957. Swallow's own contemporaneous account gives a better idea of the flavor of the work—and of physical oceanography in general:

> During the first part of the cruise . . . the weather was poor, and both ships were hove-to for two of the eight days. . . . Three neutrally-buoyant floats were used but only one gave a useful measurement. The first one was abandoned after one fix, after the *Discovery II* had collided with the anchored buoy and spent one night lying-to, with the buoy and its floats wrapped round the screw. . . . The new float could not be found at the first fix. . . .

And the one float they could track did not head south. It was only after refueling at Charleston (where Stommel left the ship) that Swallow began to have better luck. On the next leg of the cruise he released eight floats underneath and a little east of the Gulf Stream. Seven of them drifted southward. Stommel's prediction of a deep western boundary current, counter to the Gulf Stream, was dramatically vindicated.

Since then it has been upheld again and again. Deep boundary currents headed toward the equator have been found all over the world: off South America in the Atlantic, off Madagascar and along a couple of submarine ridges in the Indian Ocean, and off Japan and New Zealand in the Pacific. They have been found, in short, everywhere they have been sought, along every plausible western boundary. The ocean being the size it is, oceanographers are still finding such currents and still gauging their strength today. They are doing so with the help of far more sophisticated descendents of the Swallow float, whose progress can be tracked by satellite. They are also making use of pollutants.

Two forms of pollution pioneered in the 1950s and 1960s have turned out to have great oceanographic significance. First, until the Nuclear Test-Ban Treaty was signed in 1963, the United States and the Soviet Union conducted numerous above-ground tests of fission and fusion bombs, releasing assorted radioactive compounds into the environment. Second, during much the same period all the industrial countries seized on chlorofluorocarbons (CFCs) as a convenient way of propelling underarm deodorant and other aerosols out of the can, as well as of cooling refrigerators, automobiles, and the like. Both nuclear explosions and CFCs had their upside as far as oceanographers are concerned. The bomb blasts injected hundreds of pounds of radioactive tritium—heavy hydrogen— into the atmosphere, including an estimated 200 pounds in the peak year of 1962, the year before the test ban. With a half-life of only a dozen years, tritium is extremely rare on

Earth; the bomb tests raised its atmospheric concentration a hundredfold. CFCs, meanwhile, did not exist on Earth at all until we made them. In effect both substances entered the atmosphere suddenly, in discrete pulses, and from there they rained onto the ocean (although most of the CFCs are still in the atmosphere, eating ozone). Oceanographers have found that they can readily track the spread of those pulses, and thereby track ocean currents as well. The pollutants are a kind of dye that traces where the water goes.

Both pollutants enter the deep ocean fastest in the places where deep water forms, off Greenland and Labrador, but oceanographers were surprised to learn just how fast. By 1972, when an expedition inspired by Stommel first made systematic measurements of tritium in the Atlantic, his deep boundary current had already carried the bomb fallout as far south as 40 degrees north. Five years later, William Jenkins and Peter Rhines of Woods Hole found that the tritium had reached 30 degrees north, at a depth of 10,000 feet off South Carolina, in the area where Swallow had first measured the current. And by the early 1980s, Scripps researchers led by Ray Weiss had shown that CFCs from the Labrador Sea had crossed the equator, having traveled some 6,000 miles in 23 years.

To the layman that may sound unimpressive: the tracer measurements suggest that the deep current moves at speeds as low as an inch a second and no higher than a quarter of a knot. But until Stommel predicted the current, people doubted that even such motions were possible at crushing deep-sea pressures; and they assumed that water at the floor of the North Atlantic would have been out of contact with the atmosphere for centuries rather than decades. Moreover, the current speeds revealed by the tracers are just long-term averages. There is evidence that the deep current, along the American coast at least, is gusty.

The evidence is the abyssal storms observed by Charles Hollister of Woods Hole and his colleagues. As a graduate student of Bruce Heezen's at the Lamont laboratory in the 1960s, Hollister had helped take thousands of photographs of the ocean floor with an automatic camera. He had noticed that many of the photographs showed ripples in the soft sediment—ripples that looked to him, if not to most other oceanographers at the time, like they had been formed by strong currents. Beginning in 1979 Hollister got a chance to test his conjecture. Over the next decade he and his colleagues dropped a series of instrument packages—not only cameras but also current meters and devices for measuring suspended sediment—onto a patch of continental slope 450 miles off the New England coast, to depths as great as 16,000 feet.

The results were more dramatic than anyone expected. Again and again the camera's view was obscured by blizzards of sediment, often for days or even weeks on end; again and again, during such swirling storms, the current meter registered flow velocities as high as a knot. A knot is a lot under 16,000 feet of water: the equivalent in energy of a 35-knot wind in the atmosphere. An abyssal storm, Hollister and his colleagues estimated, could transport more than six tons of sediment per minute downstream on the boundary current. On occasion it could even knock over the camera tripod. If the camera was still standing when the dust from such a storm settled, it showed a scoured seafloor with long ripples in the direction of the current—ripples that had not been there before the storm, but that were just like the ones that had first gotten Hollister thinking nearly two decades earlier.

Even then, in truth, there had been some evidence of such intense goings-on on the seafloor; John Swallow had collected it in 1959. Having found the deep western boundary current two years before, he was looking for evidence for the other half of Stommel's abyssal circulation scheme: the flow of deep water toward the pole. That flow was supposed to be slow and subtle, so Swallow knew he would have to track his floats for a long time to see it—longer than even he might like to go to sea. He decided to move himself and his family for a year to Bermuda, which happens to lie east of the boundary current, inside the great expanse of ocean where Stommel had predicted a poleward flow. Stommel secured him the loan of a 70-foot yacht from Woods Hole. Swallow's idea was to cruise out to various spots near the island, drop clusters of floats to depths of up to 13,000 feet, and then go back home. Then a few days or even a week later he would sail out again and listen for his pinging floats. Since they would only be moving half a mile a day or less, he reasoned, they should not be hard to find. Over the course of a year, Stommel's northward flow would become apparent in the float tracks.

There was just one problem: Swallow could not track the floats, at least not at first. From one week to the next the floats would disappear. And on those occasions when he could find them, he discovered that they had been moving at velocities of five miles a day, typically, rather than a fraction of mile. Sometimes they moved as fast as half a mile an hour. Worse, they showed no discernible predilection for heading north, or any other direction in particular. They scattered toward all points on the compass.

Swallow had failed to find Stommel's poleward flow, but he had discovered something just as important: eddies. More precisely, he had discovered eddies on the "mesoscale," as oceanographers call it. Everyone knew that the ocean rotated at large and small scales—that it had quasi-permanent basin-spanning gyres, as well as transient swirls on every breaking wave. But Swallow's float tracks were the first direct measurement of rotation going on at the intermediate or mesoscale—of eddies in the ocean, long-lasting but not permanent, with diameters of roughly 10 to 100 miles. Those eddies are the physical analogue of local weather systems in the atmosphere. Swallow had discovered ocean weather.

For the next three decades the study of ocean weather became a prime focus of physical oceanographers. The abyssal storms that Hollister discovered are but one dramatic aspect of it. Just as dramatic, and readily visible on satellite images of the ocean surface, are the closed rings of warm or cold water that are constantly spinning off the Gulf Stream; they may even help trigger the abyssal disturbances by spinning the ocean right to the bottom. At intermediate depths, meanwhile, depths of two-thirds of a mile or so, the Mediterranean is constantly shooting spinning lenses of warm, salty water into the Atlantic. In 1984 a team of oceanographers dropped a latter-day Swallow float into one of these "Meddies." They had found it northwest of the Canary Islands and about 800 miles southwest of Cape St. Vincent, Portugal, which is believed to be the Meddy spawning ground. By that time the Meddy was already about two years old. But the researchers managed to track it for another two years, as it drifted nearly 700 miles south. Spinning clockwise every five or six days, it gradually shed its load of two and a half billion tons of excess salt, until finally it exhausted itself and became one with the Atlantic—with North Atlantic Deep Water in particular.

Compared with atmospheric storms, which live weeks rather than years, ocean eddies are remarkably long-lived. One of the members of the Meddy research team, Philip Richardson of Woods Hole, later compiled all the available float data from Gulf Stream rings, Meddies, and other eddies. He estimated that at any given time there were around a thousand eddies drifting around the North Atlantic, spaced on average about 120 miles apart and each about 50 miles across. The ocean, it seems, is a mosaic of eddies; and as a result when oceanographers throw a bunch of floats into a current like the Gulf Stream, one we are used to thinking of as a broad arrow sweeping across the ocean, they do not get a broad sweeping arrow made up of parallel float tracks. They get a "spaghetti diagram" of tangled, looping, crossing tracks, on which the Stream itself can barely be made out.

Yet it is there. So too is the global circulation that Stommel envisioned there, in some form or other, emerging from the tangle of eddies and gyres. Even today no one has yet managed to pick out Stommel's weak poleward flow of deep water, but it must be there in some form or other; the physics require it. And over the past decade or so the interest of many oceanographers has once again turned to this general circulation pattern. After several decades of focusing on the ocean's weather, they have realized anew the importance of its climate. They have realized anew what was first realized by Count Rumford: that the climate of the whole planet depends on ocean climate.

Stommel himself never showed great interest in the effect of ocean currents on climate. "The general circulation of the world's oceans is a matter of great interest," he wrote in 1955, right after he had first dreamed up his abyssal-circulation theory, "not only from various practical points of view—climate, fishing, dumping of radioactive wastes and so forth—but primarily from the standpoint of understanding the dynamics and history of the planet on which we live." On the other hand, toward the end of that same article in *Scientific American*, Stommel did allow himself to speculate on what might happen to climate if one were to build a dam across the Strait of Gibraltar, for the sake perhaps of bringing hydroelectric power to Spain and North Africa. (This had actually been proposed.) Damming the flow of salty Mediterranean water into the Atlantic, Stommel observed, would necessarily make Atlantic water less salty and less dense. The water off Greenland might then no longer be dense enough to sink to the ocean floor, which would completely rearrange the abyssal circulation. Without the sinking off Greenland, meanwhile, the warm waters of the Gulf Stream might no longer be drawn to far northern latitudes. The sea there might freeze over. By some argument that Stommel did not trouble to explain, and that was probably wrong, this could result in a general warming of the planet rather than a cooling.

Stommel did not really believe it all himself. "All such speculations," he concluded, "merely illustrate how little actual knowledge we have . . ." And yet a few years later he returned one more time to the climate question, this time while reporting the results of a simple thought experiment. In that little paper, not noticed much at the time but now cited with increasing frequency, Stommel imagined two reservoirs of salt water, with a common surface at the top and connected by a tube at the bottom. Density differences between the reservoirs, created by differences in their temperature and salinity, could drive water from one reservoir to the other through the bottom tube. As models of the ocean's thermohaline circulation go, it was extremely simple, which is what allowed Stommel to subject it to mathematical analysis. The analysis produced an interesting result: if the system started out with warm, salty water in one reservoir and cold, less salty water in the other—in other words, with the two density-controlling variables at loggerheads—it could settle into two different and equally stable patterns of flow. Water could flow either from cold to warm or from salty to less salty. Moreover, the flow could jump fairly easily from one flow pattern to the other.

These days the North Atlantic is warmer and saltier than the Pacific, on the whole, and deep water flows out of the Atlantic and into the Pacific. Could that thermohaline circulation be perturbed enough to jump into a different flow regime? "If so," Stommel wrote, "the system is inherently fraught with possibilities for speculation about climatic change." Oceanographers and other researchers have been doing a lot of speculating along just those lines over the past decade. They have also collected a lot of evidence—evidence that the thermohaline circulation actually has jumped from one regime to another in the past, and could do so in the future, possibly with our help. To be sure, there are still no serious plans to build a dam at Gibraltar. But we may be on our way to achieving a similar effect on the ocean by subtler and more insidious means.

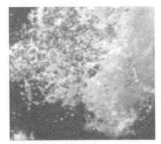

Turning Off
the Currents

W allace Broecker, or Wally to just about everybody—as in *The Glacial World According to Wally*, the title of one of his self-published books—first began to worry about global warming, at least in print, in 1975. He had been worried about the ice ages ever since his Ph.D. thesis, which he wrote at the Lamont Geological Observatory in the late 1950s. Today Broecker is still at Lamont, in a large office that also contains a lifesize Danny Devito cutout, a bearded lady mannekin, a poster of Dolly Parton reclining against a haystack, a life preserver—it is hard to know which of the myriad gewgaws merit mention as a key to Broecker's character. The Mickey Mouse Club cap, perhaps? The orangutan photograph captioned "Great thinkers like me don't need computers"? (Broecker's secretary prints out and answers his e-mail for him.) More significant, no doubt, if less unusual, is the outdated map of the ocean floor under the glass on his vast desk. Broecker started at Lamont about when Bruce Heezen and Marie Tharp were starting work on that map in the old Lamont manse overlooking the Hudson. He dates from an era when oceanography was young, and a boy could ask big questions about the ocean without huge tomes of technical literature tumbling off the shelves to crush him—questions like: What does the seafloor look like? Why is there a Gulf Stream? What causes ice ages? Broecker's adviser urged him to answer that last question in the conclusion to his thesis. "You might say I'm still writing the last chapter," Broecker says. In his seventh decade he is still boyish in other ways as well, like his unflagged irreverence and catalytic enthusiasm.

Broecker has been studying the thermohaline circulation for four decades now, except that he has a different name for it: he calls it the conveyor belt. For an article in *Natural History* in 1987, Broecker had an artist draw a picture of the conveyor belt. It showed a broad band of deep water sweeping down the center of the Atlantic to the Antarctic, spreading into the Indian and Pacific, welling up to the surface there, and returning as an equally broad and unwavering band to the North Atlantic. This image drives some physical oceanographers crazy, because they have spent the past few decades realizing just how complicated the flow of water in the ocean really is.

Much of the water that sinks in the North Atlantic never leaves that basin; it gets recirculated back to the source. The water that does get out of the Atlantic is thoroughly unrecognizable by the time it reaches the Indian Ocean, because it has been completely mixed in the Antarctic with water from different places. And the water that returns to the Atlantic does so by spinning through a gear work of eddies and gyres, not by riding a conveyor belt. The conveyor belt image is all wrong, a crude simplification at best—and of course Broecker knows that. But he also knows that, notwithstanding its wheels-within-wheels complexity, the thermohaline circulation does something very simple and important: it transports heat into the North Atlantic and salt out of it. In that sense it is not so different from a conveyor belt.

Broecker is a chemical oceanographer, not a physical one; he got interested in the thermohaline circulation when Stommel incited him and a bunch of other chemists to organize GEOSECS, the expedition that first measured the concentrations of tritium and other tracers in the ocean. He got really interested when he first realized that climate might change rapidly and drastically if the conveyor belt were ever to switch off. Broecker remembers exactly when that was. It was in 1984, in Bern, Switzerland, while he was listening to a physicist named Hans Oeschger of the University of Bern. Oeschger was lecturing on the climate record contained in a mile-and-a-quarter-deep ice core extracted from the Greenland ice sheet.

The Greenland ice sheet, like the Antarctic one, is a relic of the last ice age, and by the time of Oeschger's lecture there was a well-developed theory of what causes ice ages, the most drastic climate fluctuations of all. The theory attributed them to cyclical changes in Earth's orbit—the so-called Milankovitch cycles—which change the seasonal distribution of sunlight falling on the Northern Hemisphere. The key evidence came from the ratio of oxygen isotopes, heavy oxygen 18 and light oxygen 16, in deep-sea sediments. During an ice age, when a lot of water evaporates from the ocean and gets frozen into continental ice sheets, water containing the heavy isotope tends to remain behind. The shells of sea creatures, especially microscopic foraminifera, become enriched in it, and so do the sediments those creatures make when they die—the more ice there is on Earth, the greater the oxygen 18 enrichment in the seafloor mud. By analyzing sediment cores, oceanographers had established that ice sheets had swept over the Northern Hemisphere every 100,000 years for the past 700,000 or so, in fits and starts that were measured in tens of thousands of years. The Milankovitch theory more or less explained that pattern.

But it could not account for what Oeschger was seeing in the ice core from a site called Dye 3 in southern Greenland: evidence for far more rapid climate fluctuations during the last glaciation. One strand of evidence was, again, the ratio of oxygen isotopes. As the seafloor is getting enriched in oxygen 18 during an ice age, the ice in places like Greenland becomes depleted: the colder the air is, the less likely it is that water vapor containing the heavy isotope will make it to Greenland before precipitating out of the atmosphere. Thus the oxygen isotope ratio in the Greenland ice is a thermometer. It measures how cold the air was over Greenland when the ice was laid down.

Oeschger's second strand of evidence was actual samples of that ancient air—tiny bubbles that became trapped inside the ice when it formed. He and his colleagues had discovered they could analyze the chemical composition of those bubbles by putting a half-inch ice cube in a vacuum chamber and crushing it between beds of needles; the ancient ice would then release its breath of the past into their spectrometer. In 1982 Oeschger's group had been one of the first to report that the atmosphere during the last glaciation was different in a very important way from the preindustrial atmosphere, the one that existed right before we started aggressively burning fossil fuels: it contained only about two-thirds as much carbon dioxide. That made sense, since carbon dioxide tends to warm Earth by trapping heat. But it was not easy to see how small fluctuations in Earth's orbit could change the CO_2 level.

And the findings Oeschger reported in 1984, at the lecture Broecker heard, seemed even more distant from the Milankovitch theory. By then he and his colleagues had analyzed one section of the Dye 3 core in great detail, measuring changes over small time intervals. Ice cores permit that in a way that most sediment does not, because the ice is not riddled with worms. In most regions of the seafloor, the mud is churned and mixed to a depth of a few inches by worms and other burrowing animals. If the sediment is accumulating at a rate of less than an inch per millennium, which is the norm in the deep, the animals erase any record of climate fluctuations shorter than a few thousand years. Ice in Greenland or Antarctica, on the other hand, is laid down in neat annual layers that often remain undisturbed. In some places the layers can be counted like tree rings, thousands of years into the past.

Oeschger's group had analyzed one section of the Dye 3 core that way. The ice in that section had been laid down 40,000 to 30,000 years ago, during the height of the last glaciation. Yet, remarkably, its oxygen isotopes showed that during that period the climate had not been unwaveringly cold. Abrupt fluctuations in the isotope ratio revealed that the mean annual temperature over Greenland had risen as much as 13 degrees Fahrenheit within just a decade or two, then stayed high for a millennium before falling just as rapidly. And when the Swiss researchers popped the air bubbles in the ice, they found something more remarkable still. The carbon dioxide concentration of the ancient atmosphere seemed to have fluctuated in lockstep with air temperatures. In just a thousand years or so it had risen and fallen by as much as a quarter.

The temperature fluctuations had been seen before. Willi Dansgaard, the Danish researcher who had first suggested that ice cores would make good climate records, had found oxygen-isotope swings comparable to Oeschger's along the whole length of the Dye 3 core and in another Greenland ice core as well. Dansgaard had suggested that "shifts between two different quasi-stationary modes of atmospheric circulation" might explain the pronounced temperature shifts. But Oeschger's carbon dioxide measurements suggested something else must be going on. The atmosphere could certainly not change its own carbon dioxide concentration by 25 percent. In his talk that day in Bern, Oeschger hinted that the answer might lie in the ocean, which is a giant reservoir of dissolved carbon dioxide. At that point Broecker's mind leapt into a different quasi-stationary mode, from which it has yet to emerge. Maybe it was *ocean* circulation that was changing, he thought: "I said, oh my God, if you turned on and off the conveyor, it would do exactly what you want."

Even today no one, including Broecker, can say exactly how changes in the thermohaline circulation might have produced dramatic changes in atmospheric CO_2. And for the moment the question is moot anyway—because no one, including Oeschger, has been able to detect the rapid CO_2 fluctuations in ice cores from other regions of the world. Although no one doubts that ice age CO_2 levels were far lower, there is considerable doubt that they fluctuated dramatically; the results from Dye 3 may reflect some local environmental phenomenon. The sharp peaks and valleys in the oxygen-isotope record, on the other hand, are definitely real; they have been seen in cores from all over the world. During the last ice age the climate in the North Atlantic region really did lurch back and forth between cold and relatively warm conditions. Broecker calls these millennial lurches Dansgaard-Oeschger events. And his explanation for them, though it was inspired by Oeschger's dubious CO_2 results, has fared better than those results themselves. The conveyor belt does seem to have switched states in the past and, in so doing, to have changed the amount of heat it transports to the North Atlantic.

The best-documented case, naturally, is a relatively recent one. It was waiting ready-made for Broecker when he heard Oeschger's talk. Long before anyone dreamed of Dansgaard-Oeschger events, European paleobotantists had discovered that the final retreat of the ice sheets did not go smoothly. It started rapidly and promisingly enough, around 16,000 years ago, but then around 12,500 years ago, the temperature plummeted again. For more than a millennium, Europe was plunged back into glacial conditions. The forests that had only lately taken over the landscape gave way again to Arctic shrubs and grasses, including a wildflower, *Dryas octopetala*, that—presumably thanks to its well-preserved remains—ended up giving its name to the whole sorry period: the Younger Dryas.

The cause of the Younger Dryas had been a mystery (albeit

not one that had disturbed the rest of very many people). Expanding on an idea first suggested by the oceanographer Claes Rooth, Broecker proposed that the resurgence of cold conditions in Europe had been triggered by a collapse of the conveyor belt. During the coldest parts of the ice ages, when sea ice spread south past Iceland and over the region where deep water forms today, deep-water formation and thus the whole conveyor had been shut off. As the ice began its rapid retreat 16,000 years ago—driven ultimately by the Milankovitch variations in sunlight—warm, salty water again reached the region north of Iceland. There it gave up its heat to the cold west winds, which shipped most of it to Europe. The chilled, salty water sank to the seafloor, thus starting up the conveyor. As the conveyor transported more and more heat to the north, it accelerated the retreat of the ice.

Then something curious happened. In North America, in what is now southern Manitoba, a giant lake of glacial meltwater had formed to the west of the lobe of continental ice that protruded south into the central United States. Called Lake Agassiz, after the nineteenth-century Swiss-born naturalist who had recognized the reality of ice ages, it was larger than all the present Great Lakes combined. At first its water drained down the Mississippi into the Gulf of Mexico; in sediments there oceanographers have found its distinctive isotopic signature—depleted in oxygen 18, like all continental ice. But as the ice sheet retreated north past the Great Lakes region, a new and shorter path to the sea was opened: through the Great Lakes Basin and into the St. Lawrence. Thirty thousand tons a second of freshwater began rushing into the North Atlantic from this new source, right into the northward-bound leg of the conveyor belt. All that freshwater substantially diluted the water in the conveyor; in fact, the seawater was no longer salty enough to sink to the ocean floor by the time it reached Greenland. Without that sinking, the conveyor was shut off. So was the heat the conveyor delivers to the North

Atlantic region. The ice advanced again, and *Dryas* flowers began blooming again on the plains of northern Europe.

Sediments in the Atlantic itself record the throttling of the conveyor. Mud from the St. Lawrence, the Hudson, and other American rivers is carried south on the stormy deep-boundary current, and some of it washes up onto the slopes of the Bermuda Rise, a broad seafloor elevation 400 miles northeast of the island. Sediment accumulates there ten times faster than it does on a typical patch of seafloor—too fast even for the most enthusiastic worms to erase its record of millennial climate fluctuations. In 1987, not long after Broecker first proposed his theory, Ed Boyle of MIT and Lloyd Keigwin of Woods Hole reported that the Younger Dryas was readily discernible in that sediment. During warm periods like today, they found, mud-dwelling foraminifera on the Bermuda Rise absorb into their shells the distinctive chemical imprint of the North Atlantic Deep Water that washes over them. But during the Younger Dryas, the forams had been stamped instead by the rival water mass: Antarctic Bottom Water, invading from the south and apparently meeting little resistance. The North Atlantic Deep Water must have been weak then—which is another way of saying the conveyor belt was weak, and possibly had turned off altogether. This result was very gratifying to Wally Broecker.

It was also not so different from the result Stommel had imagined arising from a dam across the Strait of Gibraltar. Stommel had pictured a way of cutting off one of the main sources of salt to the conveyor; Broecker was instead diluting the salt with freshwater. A chance diversion of a single large river, it seemed, could drastically change the climate of a large part of the globe. Computer models of the thermohaline circulation soon confirmed that it was possible. Indeed, it did not even take a freshwater source as large as Lake Agassiz.

But the Younger Dryas was just the most recent in the long series of millennial climate swings that marked the last ice age, and that are faithfully recorded in the Greenland ice sheet. What about the others? After Boyle and Keigwin reported their results, oceanographers began finding other places in the ocean where the sediment collected fast enough—or the burrowing worms were sluggish enough—to record rapid climate fluctuations. They soon discovered that the sediment record was as spiky as the one in the ice sheet. Perhaps the most complete record was reported in 1995 by Delia Oppo and Scott Lehman of Woods Hole. In the sediment core library at Lamont, they found a core collected 30 years earlier that preserved 200,000 years of climate history in minute detail in its 30 feet of mud. The core had been extracted from the east flank of the Mid-Atlantic Ridge, due south of Iceland. Today that site is bathed in North Atlantic Deep Water cascading out of the Greenland Sea.

But that was not always the case. By picking forams out of the mud and measuring their ratio of carbon isotopes—one of the signatures of a water mass—Oppo and Lehman found that the North Atlantic Deep Water spigot had been turned down or off entirely dozens of times during the last ice age. During those intervals, apparently, Antarctic water had advanced right up to a latitude of 60 degrees north, lapping against the foot of Iceland. Judging from the sediment core, there was never any peace at all in this hundred-thousand-year north-south war of the water masses; the front surged back and forth constantly, rapidly—on the time-scale of centuries, anyway—with each shift in fortunes corresponding to a major shift in the operation of the conveyor.

All these shifts, obviously, could not be blamed on the capricious drainage of Lake Aggasiz. Nor does there seem to have been an abundant supply of other giant lakes during the last ice age waiting to be diverted at regular intervals into the North Atlantic, there to put a cap on the formation of deep water. On the other hand, there certainly was an abundant supply of ice. And other sediment cores in the Lamont archive suggest it had a dramatic influence on the operation of the conveyor belt circulation, and thus on climate.

"Sediment cores suggest . . ."—the phrase scarcely does justice to the suffering of sedimentologists and to the painstaking labor that goes into extracting even a single clue to Earth's climate history from a long column of seafloor mud. Extracting the core itself is not the half of it. During the 1950s and 1960s, Lamont scientists were under orders from the director of their laboratory, Maurice Ewing, to pull up a core every day they were at sea, wherever they might be. Today, as a result, the Lamont archive contains 18,000 cores in various states of desiccation and refrigeration.

Finding the right one for your purposes is one problem, but Gerard Bond has an advantage there; his office adjoins the core archive and his wife, Rusty Lotti, is its curator. The bigger problem is teasing climate information out of the core once you have it, with nothing to sustain you through the long hours of tedium but doggedness and faith—faith that in the end, a scattering of sand grains and microscopic shells in a petri dish may vouchsafe to you the reality of a dramatic change in Earth's climate tens of thousands of years ago. A rearrangement of ocean currents and winds, a surging of ice sheets—all this is there in a handful of sand or less, if you look closely enough. Bond and Lotti have spent years scalpeling through a few select sediment cores; Bond reckons that he personally has counted 700,000 sand grains, one by one under a microscope, sorting them by type. "No geologist in his right mind would ever do anything like this," he says, except, perhaps, a geologist who has strayed into the orbit of Wally Broecker.

Bond came late to the study of marine sediments, or at least recent ones. His career had been devoted to the study of sedimentary rocks on land, mostly half-billion-year-old Cambrian formations in the Canadian Rockies. In the late 1980s, though, he conceived the idea that he could see evidence of Milan-kovitch cycles in the shifting colors of the strata. As a way of testing that idea, he started looking at recent sediment cores, in which the evidence for Milankovitch cycles was well established. The dried-out cores themselves did not show color variations very well, but fortunately for Bond the researchers who extracted the cores had routinely photographed them while they were fresh and wet and published those photographs in books—page after page of section after section of mud. Bond cut up an article devoted to one core, called DSDP 609, and pasted the photographs end to end on the wall outside his office. He now had 700,000 years of climate history running down a 30-foot hallway. Looking at the photographs from an angle, he could readily see the sequence of ice ages and warmer interglacials marching down the hall in a kind of binary code: dark, light, dark, light, dark, light. And when he digitized the photographs and measured the core's color more precisely in 200 shades of gray, he could tell that it varied tremendously on a much more rapid timescale than that of ice age and interglacial.

Bond decided this variability was worth studying and wrote up a proposal to secure the necessary grant. He still thought of the project as little more than a brief detour out of the Cambrian Period. And he did not expect much when, as a courtesy, he sent a copy of the proposal over to Broecker, whose professional turf he was proposing to tread on. Broecker was far from resenting the intrusion. "Wally knew all about ice cores and these problems of abrupt climate change—I knew nothing about that at the time," Bond recalls. "He came tearing over to my office. He saw the gray-scale shifts and he said, 'That's just like the ice-core record.' That led to the collaboration between the two of us that has gone on for some time. So that was how I got started. Wally really twisted my arm."

By then Lamont scientists had long since figured out what the light and dark stripes in an Atlantic sediment core represented. The light sediment consisted mostly of calcareous foram shells, deposited in a period of relatively equable climate. The dark sediment, on the other hand, came from far away: it consisted of grains of rock scraped off the land by advancing ice sheets, carried out to sea by icebergs, and deposited on the ocean floor when the icebergs finally melted. No other process could explain how sand-grain-sized particles could make it to the middle of the ocean; winds or currents could not carry them so far. Thick stripes of iceberg-deposited sediment at a latitude of 50 degrees north, where Bond's DSDP 609 came from—the latitude of the south coast of England—obviously must have been deposited in periods that were pretty cold. But until Bond started quantifying the subtler color variations in his core, no one had realized that they indicated much more rapid fluctuations in climate.

With Broecker urging them on, Bond and Lotti and a couple of technicans started dissecting DSDP 609 as no core had been dissected before. They cut samples out of every other one of its 800 centimeters—out of every century and a half of climate history. Each thimbleful of mud was filtered to separate out the shells and grains of rock. Then those particles, spread into a grid of compartments on a palm-sized tray, were subjected to several stages of analysis. First one technician would pick over the sample looking for planktonic, surface-dwelling forams; if they were predominantly of a polar species whose shell coils to the left, it meant that the sea surface over the sediment core had been very cold during that period. Then another technician would go over the same sample to pick out the bottom-dwelling forams, scanning the scattered grains under a binocular microscope and gently lifting out the white, toothlike shells with the moistened tip of a fine paintbrush. It took an hour to do one sample, and after that you might end up with no forams at all; but if you had at least two or three, you could measure their oxygen- and carbon-isotope ratios. Finally Bond himself scanned the sample to sort the rock grains. Those grains could tell him, a sedimentary petrologist with decades of experience, where the icebergs that deposited them had come from. It would have taken him years to train a technician to do that reliably.

One of first things Bond noticed, when he started examining the samples from DSDP 609, was that there was something wrong with equating light sediments with forams and dark sediments with ice-rafted rock. There were places in Bond's core that were relatively light and yet foram-free—because they were crammed with grains of white limestone. "It really shocked me," Bond recalls. "You would think that with icebergs coming from all these different sources, there would be a mix of things. And the layers above and below this were the normal mix of quartz and feldspar and very minor amounts of limestone. Then all of a sudden, BOOM, there was this enormous amount of limestone, a huge change in the composition of the grain. There aren't that many places where that kind of stuff can come from."

In fact there was only one place that was plausible, one place on the North Atlantic rim where an advancing ice sheet was likely to have ground over limestone bedrock: the Hudson Strait, at the mouth of Hudson Bay in Labrador. When Bond got in touch with another researcher, John Andrews of the University of Colorado, who had analyzed sediment cores from the Labrador Sea, he learned that the limestone layers were present in those cores too, and being closer to the source, they were much thicker than the ones in DSDP 609. And from Broecker, a ravenous consumer of climate literature, Bond learned that a German oceanographer, Hartmut Heinrich, had identified the same layers a few years earlier in a core a couple of hundred miles southeast of DSDP 609.

An astonishing vision took shape in Bond's mind: a vision of a giant ice sheet surging through the Hudson Strait, its underside melting and refreezing around shattered bits of limestone; and of a vast armada of icebergs setting sail from the thunderously collapsing edge of that ice sheet, drifting down the Labrador Sea and out across the North Atlantic on the prevailing current, gradually melting and dropping limestone on their way. A couple of glaciologists later tried to estimate how much sediment might have been deposited in just one of these "Heinrich events," and they came up with a figure of around a trillion tons. Bond himself estimated how much freshwater the melting icebergs might have shed into the surface layer of the North Atlantic. He put the concentration at 1 part in 30, which is about what you would get by dropping an ice cube into every quart of ocean. That would be more than enough to freeze the conveyor belt.

Heinrich events happened every 7,000 to 10,000 years or so during the last ice age. But as Bond and Lotti tore deeper into DSDP 609 and another core from the eastern Atlantic, they began to see that Heinrich events were just the tip of the iceberg, as it were. Dense layers of dark rock grains between the Heinrich layers indicated that smaller iceberg armadas had been launched more frequently—but not from the Hudson Strait, because the grains were not limestone. Sorting them into 15 different types, Bond found that two types stood out in each dark layer: black volcanic glass from Iceland, whose active volcanoes at the time poked through a thick ice cap, and redstone from the Gulf of St. Lawrence. Iceberg fleets had apparently departed from those ports every 1,500 years, and every third or fourth one of them had encountered an even larger armada from the Hudson Strait. Nearly all the iceberg fleets coincided with Dansgaard-Oeschger events, that is, with periods of sharply cooler air over Greenland.

Every 1,500 years, then, the following events occurred in the North Atlantic region: the air over Greenland, having suddenly warmed nearly to interglacial temperatures, plunged back into deepest cold in the space of a decade. Ice sheets in North America and Iceland, and possibly elsewhere as well, discharged fleets of icebergs that drifted across the ocean and as far south as 45 degrees north. And the formation of deep water in the North Atlantic was stopped or sharply curtailed. "Sediment cores suggest," especially those analyzed by Boyle and Keigwin and Oppo and Lehman, that the conveyor belt was weakened during the last ice age but never turned off entirely. Water continued to sink in the North Atlantic, but it was apparently not salty enough to sink all the way to the bottom. It settled instead at an intermediate depth, flowing southward, with Antarctic water sloshing northward underneath it.

All these events happened repeatedly in the last ice age—but unfortunately, researchers cannot be sure in what order. When they look up from their sediment or ice cores, they are haunted by the specter of the chicken and the egg. Perhaps the ice sheets, responding to their own internal rhythm of growth and decay, launched their iceberg armadas whenever they had gotten too fat; the melting ice then clamped down on the conveyor, and the weakened conveyor transported less heat to the North Atlantic, thereby cooling the air over Greenland. But then why would at least two different ice sheets decide to purge themselves simultaneously, as Bond had discovered? Perhaps instead the air got colder first, which caused all the ice sheets around the North Atlantic to surge into the sea, which turned down the conveyor, which made things colder still. But then what cooled the atmosphere in the first place?

Add to this dilemma another one: geography. When Broecker first started thinking about millennial climate cycles, and the Younger Dryas in particular, he was looking to explain how temperatures in the North Atlantic region could ever have taken a sudden nose dive—much too sudden to have anything to do with Milankovitch cycles. Computer models of Earth's climate, chiefly the one developed by Syukuro Manabe at the Geophysical Fluid Dynamics Laboratory in Princeton, confirmed Broecker's hunch that the conveyor belt could do the job by switching abruptly to a weakened state. The models even reproduced the regional extent of the Younger Dryas cooling, which at the time was thought to have been felt primarily in Europe and to a lesser extent in eastern North America. But in the last decade the evidence has changed. The Younger Dryas and the other Dansgaard-Oeschger events are no longer merely North Atlantic curiosities. "No way can I get gigantic cooling everywhere," grumbles Manabe. Yet that is what the evidence points to, and it comes from some unusual places.

Huascarán, Peru, is not the first spot most researchers would think to look for the causes or effects of changes in the North Atlantic. It is a glacier-covered mountain in the Andes, 9 degrees south of the equator and 200 miles north of Lima. The highest of its twin peaks reaches 22,205 feet. Lonnie Thompson of Ohio State University did not make his drilling team climb that high; they stopped just shy of 20,000 feet with their six tons of equipment, at a saddle point between the two peaks, where the ice was more than 700 feet thick. Thompson has been drilling into mountain glaciers for more than two decades now, ever since he got bored with drilling in Greenland and Antarctica. He is used to skepticism from colleagues. Not long after he started on mountains, with his Ph.D. not even finished, Willi Dansgaard, the polar drilling pioneer, wrote a letter to him and to his funding agency saying that the technology did not exist to do what Thompson wanted to do. This did not help Thompson's cause. But Dansgaard was right, as Thompson already knew: he had discovered that on his first expedition in 1979, to a glacier called Quelccaya in southern Peru.

"We were naive," he recalls. "We thought we could use a twin-engine Bell 212 helicopter and bring a drill up from Antarctica, and we'd get it up there and drill the core and that would be it. But we found that, in fact, you couldn't. The elevations we work in, above 19,000 feet, are really out of the range of most helicopters, and when you have a lot of convective activity in the mountains, it makes flight very difficult and dangerous. We'd be flying along at 19,000 feet and the helicopter would just fall. The pilot's eyes were large, and I'm sure ours were also! There was no way we could get near the surface." Because the technology did not exist to land a big ice drill on an Andes peak, Thompson drew a logical conclusion: he would build a drill light enough to carry up on backs—his own, his graduate students' and those of a few dozen porters and mules. If the technology did not exist, he would invent the technology.

In 1993, 14 years after that first failure, Thompson found himself camped on Huascarán with a lightweight carbon-fiber drill and 60 solar panels to power its heated, ring-shaped tip through the ice. As each five-foot length of ice core was extracted from the borehole, it went into insulated packing material and then into a walk-in storage cave that Thompson and his crew had dug into the glacier. When the cave was full, the porters were called. Working in the pitch darkness of 3 A.M.—the coldest and so most desirable time of day—they hoisted the ice onto their backs and carried it down a 50-foot ladder that sloped across an 80-foot-deep crevasse; then on to the edge of the glacier, where mules waited to take it to the foot of the mountain, where trucks waited to take it to a fish freezer in the town of Huaraz. Some of Thompson's graduate students did not appreciate the beauty of that crevasse, which widened steadily as the expedition wore on. ("Sometimes they made career choices when they looked at the ladder," Thompson says.) Fortunately porters were plentiful. "We happened to drill this core at the height of the Shining Path guerilla activities in Peru," says Thompson. "On one side that was a problem, because there was a danger associated with that. But on the other side, we had a complete hotel to ourselves, so we could set up a laboratory. And we had all the porters we needed." Not that Thompson himself spent much time in the hotel: he camped out on Huascarán for 45 days, working sunup to sundown in winds that ripped his tents and in air that was half as thick as the comfortable fluid we breathe at sea level.

To what end this amazing effort? At Quelccaya, Thompson had ultimately gotten his cores; and in a satisfying turnabout, Dansgaard had ended up doing some of the isotope analysis. But the climate record in those cores only went back 1,500 years. They showed evidence, for instance, of intense farming before the Incas in the region around Lake Titicaca; dust from those farms was deposited downwind on Quelccaya. But pre-Inca still meant first millennium A.D. At Huascarán, the tallest mountain in the tropics, Thompson was hoping for access to a deeper past. And when he drilled his cores to bedrock there, he got it: the ice at the bottom was 20,000 years old. It had survived intact, stuck to the mountain, since the last peak of the last glaciation.

Until recently, the conventional wisdom had been that the ice age had left the tropics largely untouched. The Huascarán cores give that view the lie: the oxygen isotopes in them indicate that at the height of the glaciation the temperature on the mountain was 15 to 22 degrees Fahrenheit below what it is today. If you extrapolate that temperature down to sea level, as Thompson did, you find that the surface of the tropical Atlantic, where the snow falling on Huascarán comes from, was at least 9 degrees colder. Like the atmosphere at high latitudes, the tropical atmosphere was also much drier in the ice age: the strata from the bottom of the Huascarán cores contain 200 times more dust than falls on the mountain today. That dust did not come from farms. It was apparently blown in from Venezuela and Colombia, where vast tracts of land that are now savanna were then covered by dune fields instead.

Most surprising of all, the Younger Dryas shows up clearly in the Huascarán record. And not just the Younger Dryas, but other millennial climate events too; and not just at Huascarán: Thompson has seen them in China as well. In 1992, the year before he drilled Huascarán, he was in the Kunlun Mountains of far-western China, drilling into the Guliya ice cap at an altitude of 20,300 feet. He was looking for a history of the monsoon system, but he got a record that went all the way back to the ice age. And the oxygen isotopes in that part of the core told him that the ice age climate in western China—as in the Peruvian Andes, as in the North Atlantic—had undergone rapid oscillations.

The evidence for those oscillations now comes from all over the world. One or more of them is recorded, in one form or another, in the surging and retreating of glaciers in Chile and New Zealand, in bogs in Tierra del Fuego, in corals on Barbados, in ice cores in Antarctica, and in marine sediments on both sides of the North Pacific. Perhaps the most spectacularly precise record was unveiled in early 1996 by James Kennett of the University of California at Santa Barbara. Kennett, the marine geologist who long ago discovered evidence that glacial meltwater had been diverted from the Mississippi to the St. Lawrence during the Younger Dryas, has now seen the Younger Dryas in the Santa Barbara basin as well—the Younger Dryas and 19 of the 20 millennial cycles known from the Greenland ice cores. During the sudden warming phases of those cycles—when the North Atlantic was warm and the conveyor was at full strength—the Santa Barbara basin got warm too, judging from plankton in Kennett's sediment core. Apparently the warming signal had traveled rapidly from the Atlantic into the Pacific.

Or vice versa. The chicken-and-egg problem only gets worse as the climate change that needs explaining gets global, and as the fluctuations on different parts of the globe seem to be occurring more or less simultaneously. Simultaneously and extraordinarily rapidly: two new ice cores drilled in the early

1990s on the summit of the Greenland ice sheet show that the millennial oscillations that punctuated the last ice age were themselves punctuated by much faster flickers of the thermometer. During the warming phase of the millennial cycles, for instance, the average temperature over Greenland climbed between 9 and 18 degrees Fahrenheit in the space of a few decades. The Younger Dryas seems to have ended even faster: temperatures over Greenland rose 13 degrees and the rate of snowfall doubled in just three years, as the Earth vaulted out of the age of dusty icy dryness into the warm wet of today. Both the rapidity and the global nature of the change suggest that the atmosphere was somehow actively involved and not just responding passively to changes in the ocean; the atmosphere, after all, moves faster.

The atmosphere, the ocean, and the ice, not to mention the orbital cycles: it is worse than the chicken and the egg. "I can't tell you what's causing what," says Lloyd Keigwin, the sediment-core specialist. "I'm not a modeler. I have a really hard time grasping these things. Even the modelers have a hard time."

"This is just a model," admits Syukuro Manabe, whose model cannot explain the evidence for rapid global changes during the last glaciation and who is skeptical of that evidence. "You don't know whether it is true."

"I don't have a clear understanding of whether the conveyor is part of a broad syndrome, or whether it is the pathogen," says Scott Lehman, a sediment-core specialist who dabbles in modeling. "We're not at the point yet of being able to say much about the order of events."

"We don't know how much of the connection is through the atmosphere and how much is through changes in the thermohaline circulation," says Kennett. "It's all linked together. That's the important point."

Yet equally important, surely, is to understand what the links are. Not long ago Broecker had the glimmering of an idea.

In 1992, while he was writing the first edition of *The Glacial World According to Wally*, he had developed a severe case of writer's block as he approached the last section, in which he had hoped to set forth his grand hypothesis of what had driven all the climate change during the last glaciation. It was more or less the same last chapter he had failed to write for his Ph.D. thesis, only now the facts had gotten considerably more complicated. Broecker found he still did not have a coherent hypothesis. By 1996, though, he was groping toward one. It was inspired by the work of Lonnie Thompson on Huascarán.

That ice core offers the strongest of several strands of evidence that the tropical atmosphere was extremely dry during the ice age. Thompson and Broecker estimate it contained only 80 percent as much water vapor as it does today near the surface, and only 40 percent as much at high altitudes. Today the tropics are the planet's largest source of water vapor; it rises there off the warm sea surface and is carried by winds toward the poles. Along the way it precipitates as rain and snow, and at the same time it serves another important function: it is the most important greenhouse gas, more important even than carbon dioxide. If the water vapor concentration in the last ice age had been substantially lower, then that alone would have cooled the planet substantially.

In Broecker's hypothesis, rapid changes in the water vapor concentration, caused somehow by changes in the conveyor belt, are what produced the millennial climate cycles of the last ice age. The most likely trigger, he says, is still a shot of freshwater to the North Atlantic—just as in his original explanation for the Younger Dryas. Icebergs streaming off the Laurentide ice sheet could weaken the conveyor over the course of centuries, culminating in the massive crescendo of a Heinrich armada; but when the last berg had melted, and the atmosphere was in the coldest and driest trough of a Dansgaard-Oeschger cycle, such that not much snow was falling on the northern latitudes, then the North Atlantic would quickly grow salty again, deep water would start forming again off Greenland, and the conveyor would spring back to life. Models such as Manabe's show that the conveyor can rebound with extraordinary alacrity when it stops getting hosed with freshwater. And a hypothesis such as Broecker's explains how a sudden warming of the North Atlantic caused by a resurgence of the conveyor can propagate with extraordinary rapidity through the atmosphere to points south—to Santa Barbara, New Zealand, and the Peruvian Andes—provided that somehow the conveyor can pump water vapor back into the tropical atmosphere.

The operative word is "somehow." The equatorial ocean is a zone of major upwelling currents, which might be expected to influence the amount of water that evaporates from the sea surface and which might in turn be under the influence of the conveyor. And in the equatorial Pacific off Peru, at least, the upwelling shuts down from time to time, during the phenomenon known as El Niño. That suggests to Broecker that the tropical atmosphere may have discreet states of operation as well, like the conveyor belt, and that it might flip in response to a flip of the conveyor. But he grows a bit exasperated when he is pressed for a more precise link between the two. "The only part of the system that we know about that has multiple states is thermohaline circulation," he says. "Okay? And we know from evidence in sediment that thermohaline circulation did change. Okay? So the working hypothesis has to be that these changes in thermohaline circulation have far-reaching effects. And what I'm trying to tell you is that we don't know what the link is. What you're asking for is the big missing piece of the whole puzzle. I mean, we have every other piece in place, and we're missing a major piece."

Would that it were only one: There is also the question of how carbon dioxide fits in—after all, its concentration was radically lower during the ice age. Maybe the ice age occurred because the orbital cycles made the planet slightly drier and dustier, such that the winds blowing over Patagonia fertilized the Southern Ocean with iron, and the phytoplankton bloomed and sucked up the carbon dioxide—just as John Martin proposed. Maybe the climate change spread from the Southern Hemisphere to the Northern, instead of the other way around, and maybe the conveyor belt was throttled only after the planet had gotten cold enough for sea ice to cover the areas where deep water forms today. And maybe, somehow, a similar sequence of changes could explain even the rapid climate oscillations during the ice age. There are many maybes and many somehows in climate research, and many possible scenarios, for the past as well as the future.

In 1991, when Lonnie Thompson went back to Quelccaya, the Peruvian glacier he had first climbed 12 years earlier, he found that it was melting. There were three lakes downhill from the ice cap that had not been there before. Thompson was disappointed but not entirely surprised. In Venezuela, three glaciers have disappeared altogether since the early 1970s. The glaciers on Mount Kenya in Africa have lost two-fifths of their combined mass since the early 1960s. "It's throughout the tropics," says Thompson. "Every glacier that we have any data on shows a very rapid retreat taking place. You have to ask why that might be." It might be that the vanishing glaciers are an early sign of man-made global warming. Even a slight warming caused by the carbon dioxide we have added to the atmosphere might be enough to evaporate a lot more water off the tropical ocean; the water vapor might then amplify the warming enough to melt the ice.

A decade ago Syukuro Manabe did an experiment with his climate model that demonstrated another effect water vapor might have. Manabe allowed the concentration of carbon dioxide to keep increasing at the rate it is now, about 1 percent per year, until after 140 years its atmospheric concentration had quadrupled. From then on he let it remain constant. As Earth's temperature rose, so did the amount of water vapor in the atmosphere, and winds carried much of that water to high latitudes, where it fell as rain and snow. In Manabe's model world, the rivers of the far north—the Mackenzie, the Ob, the Yenisey—became swollen torrents emptying into the Arctic. From there the water made its way south into the Greenland Sea. By the 200th year of the simulation, something alarming had happened: the thermohaline circulation had stopped dead.

It may never happen in real life. Manabe's model may be wrong. The fragility of the thermohaline circulation, first envisioned by Henry Stommel nearly four decades ago, has been confirmed recently by several models besides Manabe's, but conceivably they could all be wrong. Even if they are right, their assumptions may not be. The carbon dioxide concentration may not quadruple over the next century and a half; Earth's fractious community of nations may yet agree on the drastic economic and technological changes needed to limit the growth of carbon emissions in the face of rapid population growth. If the CO_2 level were only to double, Manabe's model predicts that the conveyor belt would merely weaken for two or three centuries and then restore itself, much as it may have done in the Younger Dryas. The planet might escape with a modest little ice age—and in places like northern Europe that might even come as a welcome antidote to global warming. On the other hand, Manabe's model did not take into account the possibility that the Greenland ice sheet might melt in a CO_2-warmed world. Besides flooding coastal cities, that would deliver a severe freshwater kick to the conveyor belt, right in its soft spot. Manabe says he cannot exclude the "drop-dead scenario."

So many maybes: Maybe we will shrink from rearranging the ocean currents. Maybe some marvelous technical developments will make our fears about climate obsolete. Maybe a century from now, our fretting about carbon pollution will cause our great-grandchildren to chuckle, as we do now when we read the forecasts, made before the rise of the motor car, of city streets buried in horse manure. We do not know. We do not know what will happen to ocean circulation as a result of man-made climate change; we do not know, in turn, what a collapse of the circulation would do to climate. But there is a bottom line: we do know, both from the models and from the sedimentary record, that abrupt changes in climate, probably linked to changes in the conveyor, can happen even when the planet is not in the throes of an ice age.

In a sediment core from the Antarctic, for instance, James Kennett has found signs of a sudden and dramatic change in thermohaline circulation that occurred 57 million years ago, near the end of the Paleocene, when the Earth was warmer than it is today. In the space of just two millennia the pattern of the circulation reversed: the deep ocean became warmer than the surface, as it filled with warm, salty water from the tropics instead of cold water from the poles. Such an ocean seems almost inconceivable today, but there is nothing in the physics of ocean currents to prevent it. One result of the Paleocene shift was a mass extinction among bottom-dwelling forams. It must have had much larger effects on the life of the planet as well.

Foram extinctions in the Paleocene—that is an abstract concern. Our real concern is the Holocene, the warm period we are in now, which began after the last glaciation; the period in which all human history, from the first farms and cities to the moon landings and cloned sheep, has unfolded. All that history had been thought to have taken place against the backdrop of a stable climate. But the Holocene climate, it has lately become clear, has not been stable at all. It has shown sudden variations of its own. The most recent of them may be familiar: called the Little Ice Age, it lasted roughly from the fifteenth century to the end of the eighteenth. It was the period when the Thames froze in London, the canals froze in Amsterdam, and Washington and his troops suffered at Valley Forge. Lonnie Thompson has lately seen evidence of the Little Ice Age in the ice on Huascarán; Lloyd Keigwin has seen it in sediments on the Bermuda Rise.

Apparently the Little Ice Age and other Holocene climate fluctuations had something to do with the collapse or weakening of the conveyor belt, just like the fluctuations in the big ice age. Gerard Bond has found that his North Atlantic iceberg cycles continued in attenuated form, with smaller iceberg fleets, from the glacial period right into the Holocene. One of those cool periods in the North Atlantic occurred 4,200 years ago. It seems to have coincided with a centuries-long drought in the Middle East; the evidence is an accumulation of wind-blown dust in a sediment core from the Gulf of Oman that has been analyzed by another Lamont scientist, Peter deMenocal. The drought, in turn, coincided with the collapse of the Akkadian Empire, an early Mesopotamian civilization. Maybe one collapse caused the other: maybe a change in ocean currents brought down an empire.

Of course, we have come a long way since then.

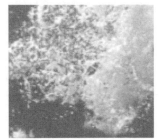

Epilogue:
An End

Water is not unique to our planet. It is turning up everywhere these days. An orbiting telescope called the Infrared Space Observatory, launched in 1995 by the European Space Agency, has confirmed that water is abundant in space, notably around dying stars that resemble our own sun. As such stars exhaust their stores of hydrogen and helium, they do not explode the way a large star would; they simply swell to an outlandish size—they become red giants—and in the process blow winds of oxygen-rich gas into space. Water forms at the shock front where this gas collides with interstellar hydrogen. That same water may later act as a sort of midwife: By carrying heat out of the interstellar cloud—in the form of the infrared radiation that the European satellite detects— water helps the cloud collapse and become the core of a new star. Without water, certainly, there would be no life on Earth; but perhaps there would not even be a solar system.

Planets are not unique to our sun. When astronomers model the process of star formation mathematically, on a computer, they find it is normal for a newborn star to be girdled by a disk of dust and gas; normal, too, for some of that dust and gas to coalesce into planets. Just in the past few years, though, they have begun to move beyond such well-founded expectations to actually discovering planets around distant stars. The planets are too dim to be seen directly, but their gravity pulls the parent star back and forth, and that regular motion leaves its mark on the starlight. By early 1998 planets had been discovered around at least eight other stars like the sun. All those planets are at least half the size of Jupiter; the instruments are not sensitive enough yet to detect something as small as Earth. A few of the planets are in dizzyingly tight orbits, revolving around the parent star in as little as four days. On a giant gas ball that close to a sun, there would be no chance of finding an ocean, and little chance of finding life.

Yet life is not likely to be unique to our solar system. The first extrasolar planets to be discovered may look bizarrely different from our own, but with 100 to 200 billion stars in our Milky Way galaxy alone, there are likely to be quite a few Earths and quite a few oceans out there. The notion that our world is special has an enduring appeal but no solid foundation. Even the new solar systems that have already been found may have, if not Earths, then worlds that are wet and habitable. Researchers at Penn State University, including James Kasting, have pointed out that at least two of the new planets orbit at promising distances from their suns; while the planets themselves are doubtless gaseous, they probably have moons that are solid. If those planets formed farther away from their star and then spiraled into their present orbits—as one current theory has it—then their moons might easily have collected enough water to make an ocean.

At least one such habitable moon may exist in our own solar system. Europa, the second large moon out from Jupiter, first glimpsed in 1610 by Galileo Galilei, is covered with water ice. It has long been speculated that there might be a liquid ocean underneath; the tidal pull of Jupiter's gravity may keep Europa warm enough. In 1997 the *Galileo* spacecraft, in orbit around Jupiter, sent back the best evidence yet. Its photographs of Europa showed regions that seemed to be a patchwork of ice blocks that had melted, drifted around a bit, and then frozen into new positions. Pack ice on Antarctic seas behaves that way. NASA has announced plans to launch another spacecraft in 2003 to map the thickness of Europa's ice and to try to verify whether it does indeed have the solar system's second water ocean. There is even talk of sending a robotic submarine to tunnel through the ice crust and nose around the ocean below. It is not out of the question that there could be hot springs on the floor of that ocean; not out of the question that those springs could be home to some primitive form of life. It would be an explorer's dream.

But for now this planet, this ocean, this life are all we have.

So limitless it seems from the beach, so limitless it truly is as a field of exploration, the ocean nonetheless has limits, as we are finding today, as exploration has been joined, increasingly over the past century, by exploitation, and as our capacity to alter nature has become global. The most fundamental limit of all, though, has nothing to do with us. The ocean will outlive us and our intrusions, but it will not live forever. Like Earth as a whole, it was born at a certain time and it will die at one.

Its impermanence is rooted in the sun, which is forever getting brighter; stars do that as they move through middle age. In seven billion years or so, the sun will become a red giant. Before that, as it swells and reddens, icy moons in the outer solar system may thaw. Besides Europa, Saturn's large moon Titan is an interesting case; it has an atmosphere somewhat denser than Earth's, one that is also mostly nitrogen but with a generous amount of methane mixed in. As the sun becomes brighter and redder, there will come a time, perhaps six billion years from now, when so much sunlight is penetrating Titan's atmosphere and becoming trapped in its methane greenhouse that its icy surface will melt—or so a pair of astronomers at the University of Arizona, Ralph Lorenz and Jonathan Lunine, have recently calculated. Ammonia ice will melt first, they say, and it will act as a kind of antifreeze that will allow water ice to melt at a temperature of −143 degrees Fahrenheit. In that frigid floor-cleanser ocean, life—a different kind from Earth's, one to which a water-ammonia mixture would not be toxic—just might begin and evolve. But Titan's moment will be briefer than ours. After a mere half-billion years, as the sun begins to eject its own atmosphere in high-speed winds, it will blow the roof off Titan's greenhouse.

At around that time Earth will probably cease to exist. When the sun swells into a red giant, it may stop just short of our orbit, but more likely it will engulf our planet and vaporize it. Life will have been extinguished much earlier, though. A few years ago Kasting and his colleague Ken Caldeira at Penn State tried to calculate exactly when—to envision the end as best as science will allow. There is no telling, of course, when or if a giant comet might smack us and sterilize the planet prematurely. Barring that bit of bad luck, however, the long-range enemies of life on Earth will be heat and carbon dioxide—or rather a lack of carbon dioxide. Just now we are worried about having too much of that gas in the atmosphere, but that is merely a short-term problem of our own creation. Over geologic time Earth has a way of getting rid of atmospheric carbon dioxide: it dissolves in rainwater and gnaws at rocks, becomes carbonate, runs in rivers to the sea, and sinks to the seafloor. As the sun gets hotter, more rain will fall, and more rock weathering will occur. Half a billion years from now the carbon dioxide concentration will be too low for most plants to continue photosynthesizing.

For as much as a billion years after that, a few hardy grasses and shrubs, plants that are especially frugal with their carbon dioxide, will soldier on. The land may resemble parts of the Australian Outback, scattered vegetation poking out of red dirt. But by a billion and a half years from now, in Kasting and Caldeira's scenario, the planet's average temperature will be about 120 degrees Fahrenheit—hot enough to extinguish everything but microbes, and rapidly getting hotter. Water, sparsely inhabited, will still cover the other seven-tenths of Earth. Life began in water four billion years ago, and many researchers these days believe bacteria at seafloor hot springs were among the first organisms. If we ever explore an ocean on Europa, that is the sort of life we would look for. Such microbes may well be the last survivors on Earth. Ultimately even the ocean will become too hot for anything else.

Over the next billion years after the extinction of plants, the atmosphere will fill with steam, as it did at the birth of the planet. But now the steam will be bombarded by sunlight that is far more intense. The ultraviolet part of that light will split water molecules in the stratosphere into hydrogen and oxygen atoms, and the hydrogen will float into space, never to come back. In space, it will meet other oxygen, oxygen cast off by the dying sun, and together perhaps they will form water on some other planet, around some other star. Two and a half billion years from now the last water molecule will have gone from Earth. From space the sea came; to space it will return. Even the seafloor microbes will expire then. From dust the planet was made; dust it will be again. For several billion years until its final incineration, Earth will be a hot, dead world like Venus, a planet too close to its sun. For the first time in its history, it will be dry.

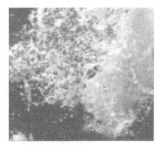

Selected
References

Most of the information in this book came from my interviews with oceanographers and from their articles in technical journals. In several historical passages I am particularly indebted to the works of Susan Schlee and Margaret Deacon. Here is a selection of sources that document some of the main points in the book, and that might be of use to someone who wants to go deeper.

Prologue: The Sea at Dawn

Carson, Rachel. *The Sea Around Us.* New York: Oxford University Press, 1951.

Dayton, Paul K. "Reversal of the Burden of Proof in Fisheries Management." *Science,* 279 (1998) 821–822.

Chapter 1: Beginnings

Bartusiak, Marcia. *Thursday's Universe: A Report from the Frontier on the Origin, Nature, and Destiny of the Universe.* Redmond, Washington: Tempus Books, 1988.

Kasting, James F., Owen B. Toon, and James B. Pollack. "How Climate Evolved on the Terrestrial Planets." *Scientific American,* February 1988: 90–97.

Matsui, Takafumi and Yutaka Abe. "Evolution of an impact-induced atmosphere and magma ocean on the accreting Earth." *Nature,* 319 (1986): 303–305.

Matsui, Takafumi and Yutaka Abe. "Impact-induced atmospheres and oceans on Earth and Venus." *Nature,* 322 (1986): 526–528.

Schopf, J. William. "Microfossils of the Early Archean Apex Chert: New Evidence for the Antiquity of Life." *Science,* 260 (1993): 640–646

Thomson, Thomas. *The History of Chemistry.* New York: Arno Press, 1975. First edition 1830.

Wetherill, George W. "Formation of the Earth." *Annual Review of Earth and Planetary S.cience,* 18 (1990): 205–256.

Chapter 2: The Seafloor Moves

Allègre, Claude. *The Behavior of the Earth: Continental and Seafloor Mobility.* Cambridge, Massachusetts: Harvard University Press, 1988.

Deacon, Margaret. *Scientists and the Sea, 1650–1900: A Study of Marine Science,* New York: Academic Press, 1971.

Drubba, H. and H. H. Rust. "On the First Echo-Sounding Experiment." *Annals of Science,* 10 (1954): 28–32.

Fox, Paul J. "Bruce C. Heezen: A Profile." *Oceanus,* 34 (1991) Number 4: 100–107.

Heezen, Bruce C. "The Rift in the Ocean Floor." *Scientific American,* October 1960: 98–110.

Hess, H. H. "History of Ocean Basins." In *Petrologic Studies—A Volume in Honor of A. F. Buddington.* Geological Society of America, 1962, 599–620.

Le Pichon, Xavier. "Introduction to the publication of the extended outline of Jason Morgan's April 17, 1967 American Geophysical Union Paper on 'Rises, Trenches, Great Faults and Crustal Blocks.'" *Tectonophysics,* 187 (1991): 1–22.

Maury, Matthew Fontaine. *The Physical Geography of the Sea, and Its Meteorology.* Edited by John Leighly. Cambridge Massachusetts: Belknap Press of Harvard University Press, 1963. First published 1855.

Menard, H. W. *The Ocean of Truth: A Personal History of Global Tectonics.* Princeton: Princeton University Press, 1986.

Schlee, Susan. *The Edge of an Unfamiliar World: A History of Oceanography.* New York: Dutton, 1973.

Tharp, Marie and Henry Frankel. "Mappers of the Deep." *Natural History,* October 1986: 48–62.

Chapter 3: To Map Is to Know

Heezen, Bruce C. "The Origin of Submarine Canyons." *Scientific American,* August 1956: 36–41.

Macdonald, Ken C. et al. "It's Only Topography." *GSA Today,* 3-1 (January 1993): 1, 24–25; and 3-2 (February 1993): 29–31, 34–35.

Macdonald, Ken C. et al. "Volcanic growth faults and the origin of Pacific abyssal hills." *Nature,* 380 (1996): 125–129.

Macdonald, Kenneth C. and Paul J. Fox. "The Mid-Ocean Ridge." *Scientific American,* June 1990: 72–79.

Oceanus, 34 (1991) Number 4: Mid-Ocean Ridges.

Oceanus, 35 (1992) Number 4: Marine Geology and Geophysics.

Smith, Walter H. F. and David T. Sandwell. "Global Sea Floor Topography from Satellite Altimetry and Ship Depth Soundings." *Science,* 277 (1997): 1956–1962.

Tharp, Marie. "Mapping the Ocean Floor—1947 to 1977." In *The Ocean Floor.* Edited by R. A. Scrutton and M. Talwani. New York: John Wiley and Sons Ltd., 1982, 19–31.

Chapter 4: Islands in the Deep

Gage, J. D. and P. A. Tyler. *Deep-Sea Biology: A Natural History of Organisms at the Deep-Sea Floor,* Cambridge, England: Cambridge University Press, 1991.

Grassle, J. Frederick. "Species Diversity in Deep-Sea Communities." *Trends in Ecology and Evolution,* 4 (1989): 12–15.

Grassle, J. Frederick and Nancy J. Maciolek. "Deep-Sea Species Richness: Regional and Local Diversity Estimates from Quantitative Bottom Samples." *The American Naturalist,* 139 (1992): 313–341.

Heezen, Bruce C. and Charles D. Hollister. *The Face of the Deep.* New York: Oxford University Press, 1971.

Levin, Lisa A. and Cynthia L. Thomas. "The ecology of xenophyophores (Protista) on eastern Pacific seamounts." *Deep-Sea Research,* 35 (1988): 2003–2027.

Mosely, H. N. "Deep-sea Dredging and Life in the Deep Sea." *Nature,* 21 (1880): 543–547, 569–572, 591–593.

Priede, I. G. et al. "Direct measurement of active dispersal of food-falls by deep-sea demersal fishes." *Nature*, 351 (1991): 647–649.

Report on the scientific results of the voyage of H. M. S. Challenger during the years 1873–76. Narrative—Vol. I London, 1885.

Smith, Craig R., Peter A. Jumars, and David J. Demaster. "*In situ* studies of megafaunal mounds indicate rapid sediment turnover and community response at the deep-sea floor." *Nature*, 323 (1986): 251–253.

Chapter 5: Springtime

Cavanaugh, Colleen M. et al. "Prokaryotic Cells in the Hydrothermal Vent Tube Worm *Riftia pachyptila* Jones: Possible Chemoautotrophic Symbionts." *Science*, 213 (1981): 340–342.

Corliss, John B. et al. "Submarine Thermal Springs on the Galápagos Rift." *Science*, 203 (1979): 1073–1083.

Edmond, John M. and Karen Von Damm. "Hot Springs on the Ocean Floor." *Scientific American*, April 1983: 78–93.

Hessler, Robert, Peter Lonsdale, and James Hawkins. "Patterns on the Ocean Floor." *New Scientist*, 117 (24 March 1988): 47–51.

Kaharl, Victoria A. *Water Baby: The Story of Alvin*. New York: Oxford University Press, 1990.

Kunzig, Robert. "Between Home and the Abyss." *Discover*, December 1993: 66–75.

Lutz, Richard A. and Rachel M. Haymon. "Rebirth of a Deep-sea Vent." *National Geographic*, 186 (November 1994): 114–126.

Mullineaux, L. S., P. H. Wiebe, and E. T. Baker. "Larvae of benthic invertebrates in hydrothermal vent plumes over Juan de Fuca Ridge." *Marine Biology*, 122 (1995): 585–596.

Turner, Ruth D. "Wood-Boring Bivalves, Opportunistic Species in the Deep Sea." *Science*, 180 (1973): 1377–1379.

Van Dover, Cindy Lee. *The Octopus's Garden: Hydrothermal Vents and Other Mysteries of the Deep Sea*. Reading, Massachusetts: Addison-Wesley, 1996.

Chapter 6: Blue Water

Alldredge, Alice. "Appendicularians." *Scientific American*, July 1976: 94–102.

Gilmer, Ronald W. "Free-floating Mucus Webs: A Novel Feeding Adaptation for the Open Ocean." *Science,* 176 (1972): 1239–1240.

Haeckel, Ernst. "Plankton-Studien." *Jenaisches Zeitschrift,* 18 (1890–1891): 232–336.

Hamner, William M. "Blue-Water Plankton." *National Geographic,* 146 (October 1974): 530–545.

Hamner, W. M. et al. "Underwater observations of gelatinous zooplankton: Sampling problems, feeding biology, and behavior." *Limnology and Oceanography,* 20 (1975): 907–917.

Harbison, G. R., L. P. Madin, and N. R. Swanberg. "On the natural history and distribution of oceanic ctenophores." *Deep-Sea Research,* 25 (1978): 233–256.

Harbison, G. Richard. "The Gelatinous Inhabitants of the Ocean Interior." *Oceanus,* 35 (1992) Number 3: 18–23.

Harbison, G. Richard. "The Structure of Planktonic Communities." In *Oceanography: The Present and Future.* Edited by Peter G. Brewer. New York: Springer-Verlag, 1983, 18–33.

Hensen, Victor. *Die Plankton-Expedition und Haeckel's Darwinismus; über einige Aufgaben und Ziele der beschreibenden Naturwissenschaften.* Kiel: Lipsius & Tischer, 1891.

Madin, Katherine A. C. and Laurence P. Madin. "Sex (and Asex) in the Jellies." *Oceanus,* 34 (1991) Number 3: 27–35.

Madin, L. P. "Aspects of jet propulsion in salps." *Canadian Journal of Zoology,* 68 (1990): 765–777.

Chapter 7: Invisible Garden

Anderson, Donald M. "Turning back the harmful red tide." *Nature,* 388 (1997): 513–514.

Chisholm, Sallie W. "What Limits Phytoplankton Growth?" *Oceanus,* 35 (1992) Number 3: 36–46.

Chisholm, Sallie W. et al. "A novel free-living prochlorophyte abundant in the oceanic euphotic zone." *Nature,* 334 (1988): 340–343.

Coale, Kenneth H. et al. "A massive phytoplankton bloom induced by an ecosystem-scale iron fertilization experiment in the equatorial Pacific

Ocean." *Nature*, 383 (1996): 495–501. See related articles on pages 475–476 and 508–517.

Hardy, Alister. *Great Waters: A Voyage of Natural History to Study Whales, Plankton, and the Waters of the Southern Ocean.* New York: Harper & Row, 1967.

de la Mare, William K. "Abrupt mid-twentieth-century decline in Antarctic sea-ice extent from whaling records." *Nature*, 389 (1997): 57–60.

Martin, John H. "Glacial-Interglacial CO_2 Change: The Iron Hypothesis." *Paleoceanography*, 5 (1990): 1–13.

Martin, John H., Steve E. Fitzwater, and R. Michael Gordon. "Iron deficiency limits phytoplankton growth in Antarctic waters." *Global Biogeochemical Cycles*, 4 (1990): 5–12.

Chapter 8: Twilight of the Cod

Brawn, Vivien M. "Reproductive Behaviour of the Cod (*Gadus Callarias L.*)." *Behaviour*, 18 (1961): 177–198.

Cushing, D. H. *The Provident Sea.* Cambridge: Cambridge University Press, 1988.

Hutchings, Jeffrey A. and Ransom A. Myers. "What can be learned from the collapse of a renewable resource? Atlantic cod, *Gadus morhua*, of Newfoundland and Labrador." *Canadian Journal of Fisheries and Aquatic Sciences*, 51 (1994): 2126–2146.

Jensen, Albert C. *The Cod.* New York: Thomas Y. Crowell Company, 1972.

Lough, R. Gregory et al. "Ecology and distribution of juvenile cod and haddock in relation to sediment type and bottom currents on eastern Georges Bank." *Marine Ecology Progress Series*, 56 (1989): 1–12.

Pauly, Daniel et al. "Fishing Down Marine Food Webs." *Science*, 279 (1998): 860–863.

Rose, George A. "Cod spawning on a migration highway in the northwest Atlantic." *Nature*, 366 (1993): 458–461.

Serchuk, Fredric M. and Susan E. Wigley. "Assessment and Management of the Georges Bank Cod Fishery: An Historical Review and Evaluation." *Journal of Northwest Atlantic Fisheries Science*, 13 (1992): 25–52.

Warner, William W. *Distant Water: The Fate of the North Atlantic Fisherman.* Boston: Little, Brown, 1983.

Chapter 9: Where the Water Goes

Armi, Laurence et al. "Two Years in the Life of a Mediterranean Salt Lens." *Journal of Physical Oceanography,* 19 (1989): 354–370.

Brown, Sanborn C. *Benjamin Thompson, Count Rumford.* Cambridge, Massachusetts: MIT Press, 1979.

Gordon, Arnold L. "Interocean Exchange of Thermocline Water." *Journal of Geophysical Research,* 91 (1986): 5037–5046.

Gordon, Arnold I. and Rana A. Fine. "Pathways of water between the Pacific and Indian oceans in the Indonesian seas." *Nature,* 379 (1996): 146–149.

Huyghe, Patrick. "The Storm Down Below." *Discover,* November 1990: 70–75.

Jenkins, William J. and Peter B. Rhines. "Tritium in the deep North Atlantic Ocean." *Nature,* 286 (1980): 877–880.

Rumford, Benjamin Thompson, Count. "Of the Propagation of Heat in Fluids." In *Collected Works of Count Rumford.* Edited by Sanborn C. Brown. Cambridge, Massachusetts: Belknap Press of Harvard University Press, 1968.

Stommel, Henry. "The Westward Intensification of Wind-driven Ocean Currents." *Transactions of the American Geophysical Union,* 29 (1948): 202–206.

Stommel, Henry. "The Circulation of the Abyss." *Scientific American,* July 1958: 85–90.

Stommel, Henry M. "Autobiography: The Sea of the Beholder." In *Collected Works of Henry M. Stommel, Volume I.* Edited by Nelson G. Hogg and Rui Xin Huang. Boston, Massachusetts: American Meteorological Society, 1995.

Weiss, R. F. et al. "Atmospheric chlorofluoromethanes in the deep equatorial Atlantic." *Nature,* 314 (1985): 608–610.

Whitehead, John A. "Giant Ocean Cataracts." *Scientific American,* February 1989: 50–57.

Chapter 10: Turning Off the Currents

Behl, Richard J. and James P. Kennett. "Brief interstadial events in the Santa Barbara basin, NE Pacific, during the past 60 kyr." *Nature,* 379 (1996): 243–246.

Bond, Gerard C. and Rusty Lotti. "Iceberg Discharges into the North Atlantic on Millennial Time Scales During the Last Glaciation." *Science,* 267 (1995): 1005–1010.

Broecker, Wallace S. "The Great Ocean Conveyor." *Oceanography,* 4-2 (1991): 79–89.

Broecker, W. S. *The Glacial World According to Wally,* Palisades, New York: Eldigio Press, 1993.

Manabe, Syukuro and Ronald J. Stouffer. "Simulation of abrupt climate change induced by freshwater input to the North Atlantic Ocean." *Nature,* 378 (1995): 165–167.

Oceanus, 37 (1994) Number 1: Atlantic Ocean Circulation.

Taylor, K. C. et al. "The 'flickering switch' of late Pleistocene climate change." *Nature,* 361 (1993): 432–436.

Thompson L. G. et al. "Late Glacial Stage and Holocene Tropical Ice Core Records from Huascarán, Peru." *Science,* 269 (1995): 46–50.

Epilogue: An End

Caldeira, Ken and James F. Kasting. "The life span of the biosphere revisited." *Nature,* 360 (1992): 721–723.

Lorenz, Ralph D., Jonathan L. Lunine, and Christopher P. McKay. "Titan under a red giant sun: A new kind of 'habitable' moon." *Geophysical Research Letters,* 24 (1997): 2905–2908.

Williams, Darren M., James F. Kasting, and Richard A. Wade. "Habitable moons around extrasolar giant planets." *Nature,* 385 (1997): 234–236.

Index